PLANT LIFE

PLANT LIFE

The Entangled Politics of Afforestation

ROSETTA S. ELKIN

UNIVERSITY OF MINNESOTA PRESS

MINNEAPOLIS • LONDON

The University of Minnesota Press gratefully acknowledges the financial assistance provided for the publication of this book by McGill University and the Harvard University Graduate School of Design.

This book is supported by a grant from the Graham Foundation for Advanced Studies in the Fine Arts.

Published by the University of Minnesota Press
111 Third Avenue South, Suite 290
Minneapolis, MN 55401-2520
http://www.upress.umn.edu

ISBN 978-1-5179-1261-1 (hc)
ISBN 978-1-5179-1262-8 (pb)

A Cataloging-in-Publication record for this title is available from the Library of Congress.

Printed in the United States of America on acid-free paper

31 30 29 28 27 26 25 24 23 22   10 9 8 7 6 5 4 3 2 1

Contents

Dry·land of, relating to, or being a relatively arid region
Af·forestation tree planting in dryland biomes

Afforestation programs incentivize planting across drylands. In the deserts of Inner Mongolia, the image of a tree such as *Populus* spp. conceals the environmental cost of planting trees, an entanglement between imported units and existing plant life. Photograph copyright by Ian Teh.

Preface

This research emerges from a particularly personal experience that I was unable to process theoretically, experientially, or spatially for a considerable time. I was living in China and visiting the landscapes that were generally peripheral to cities in order to get a sense of both the urban densities I was experiencing and their geographic context. I did so as a curious landscape architect working with the often difficult realities of extreme, unruly urbanization. This included the aberration of rescinded species catalogs, abbreviated indices of organic material, and the artifice of constructed soil while tallying specifications that ran into the hundreds of thousands of planting units. Such projects took only human relationships seriously, counting on purely economic understandings of plants as material supply. Value was determined by use, or usefulness to the growing urbanity, and thus to the disappearance of the systems that were in place before the construction, the sprawl, and the high-yield crop. In this context, physically being outside was the only means I had to engage with the ecology and the plants that had attracted me to landscape practice in the first place. But what I was experiencing and what so many others have faced is much more than a simple binary between the urban and everything else. I was caught in the global vision of Latin binomials, plant units, and engineered soils while being entirely dependent on local botanical knowledge. Each reductive system and best practice offered only a diminished familiarity and solidified my detachment from the land that was, in principle, central to my interests. I was cultivating relationships with the lists, farms, offices, and nurseries of globally industrialized trade rather than a relationship to the complexity of plant life central to understanding landscape.

A trip to the northwestern periphery of Beijing with a group of dynamic colleagues revealed the physical consequences of my concern. I took in a view that embodied the contracted scope I was coming to terms with personally and professionally. In that moment, I was looking across the landscape at a perfect grid of small plants that seemed to have been blanketed or draped across the curves of each contour. The planting was evenly spaced, without any acknowledgment of aspect, slope, or species. The territory appeared desiccated as far as we could see, and the grid spread across its vast range, re-marking and re-scripting the land. It was neither an agricultural scape nor a forestry plantation. It felt wholly unrelated to any known topologies of conservation or ecological restoration that were familiar to me, exposing a seemingly politically neutral scene.

All I could think of at the time were the commands that I was also rehearsing in practice, which muted the living plant organism, copy-paste-offset, copy-paste-offset, copy-paste-offset . . .

Many years later I learned about China's Three Norths Shelter System, a state-financed afforestation campaign. In examining some of the arguments and ideas behind this version of nature, it was clear that the project I witnessed was not a haltingly unprecedented operation or an anomaly. It was also not unique. I could not explain it ecologically, yet it was considered an environmental practice. Curiously, it dovetailed with colonial and imperial knowledge, falling under the terms of landscape *improvement,* but seemed to be described as inevitable through the lens of forestry. At the time, I turned to Dr. Richard T. T. Forman, a renowned landscape ecologist, who had also witnessed the scale of the Three Norths program firsthand. Our conversations suggested that it is as difficult to sustain a view of drylands as an independent state as it is to suggest that nature is separate from humans. He posed a question that inspires this book: Why *not* plant trees in drylands?

Over ten years have passed since I first witnessed afforestation at a scale that made me uncomfortable, and the passage of time has left its mark. The reason for revisiting this experience is to ask questions of spatial disciplines such as landscape architecture, landscape planning, and conservation, as practices are uncomfortably wedged between the practical and the scientific. They are also now uncomfortably positioned within climate change concerns that position design as a solution to a problem, without considering that spatial disciplines are often part of the problem.

The present work owes much to the time I spent in scholarly pursuits and while practicing as a landscape architect, witnessing the ramifications of a globalized profession. The design fields traditionally privilege scientific expertise, typically in the form of an elite, engineered determinism. The less spectacular knowledge, ranging from land-based practices, seed collecting, indigenous familiarity, and intuitive experiences where gravity and soils reign, are pacified by the acceleration of disembodied practices. I bring this up because I believe in practices more than professionalization, yet design fields are tending toward the latter. I look carefully at afforestation as an exploit because it relies on construction typologies, borrows from ecological rhetoric, and is ultimately a horticultural exercise. Horticulture is the culture of plants, the art or practice of cultivation and management. The projects described in this book are among the largest horticultural undertakings in the world.

Methodologically, I am dedicated to encouraging another reading of the case study, one that does not pay specific attention to historical heroes, varying levels of comparison, or the formulation of conclusive evidence. I could not possibly have entertained fieldwork in such vastly differentiated traditions, although I have witnessed fragments of each project across all three continents. Afforestation is so common and each undertaking is so vast that many readers will have firsthand experience, infusing a closer reading. I ask questions of methodology because a reorganization between design practices and practical knowledge might reveal novel cultural outcomes.

This book is neither comprehensive of the plants of afforestation nor comprehensive of the projects of afforestation. Rather, I explore how the complicated ties and relations made in the human world—namely between constructing a scientific truth, generating a resource, and misleading public understanding—are physically manifest in the landscape. I am interested in how design itself is reliant on linear scale and replication of technique, which leads me back to the experiences that instigated this research. Standing adjacent to a project of afforestation, the eye is led across the grid in such a way that it feels entirely designed. The landscape is made rigid, predictable, and static. What I am interested in is a practice that is less reliant on scalability and unit-based valuation, instead paying close attention to the landscape as a living organism because the land is alive with millions of unnamed species and unknown relationships that surpass the human imagination. This is not to say that we should not organize planting projects, plant trees, advance conservation, or plant in drylands. It is the means and methods of our plans that require our attentive notice, now more than ever.

Abbreviations

AFA	American Forestry Association
ANR	agriculture and natural resources
AOF	L'Afrique Occidentale Française (French West Africa)
AU	African Union
BLM	[United States] Bureau of Land Management
CARE	Cooperative for Assistance and Relief Everywhere
CCC	[United States] Civilian Conservation Corps
CEN-SAD	[Africa] Community of the Sahel-Saharan States
CILSS	[Africa] Permanent Interstate Committee for Drought Control in the Sahel
COP	Conference of Parties to the Climate Change Convention
CRS	Catholic Relief Services
CSA	climate-smart agriculture
EPA	[United States] Environmental Protection Agency
FAO	Food and Agriculture Organization of the United Nations
FOCAC	Forum on China–Africa Cooperation
GGW	[Africa] Great Green Wall
GGWSSI	[Africa] Great Green Wall for the Sahara and the Sahel Initiative
MDG	millennium development goal
MDP	Medium Depth Planter
NEPAD	[Africa] New Partnership for Africa's Development
NWH	[China] Northwest Half of Hu Line
PAGGW	[Africa] Pan-African Agency of the Great Green Wall
PCA	Plant Conservation Alliance
REAAP	Resilience through Enhanced Adaptation, Action-learning, and Partnership Activity

SEH   [China] Southeast Half of Hu Line
SFA   [China] State Forestry Administration
TNAP   [China] Three-North Afforestation Project
TNSS   [China] Three Norths Shelter System
UNCCD   United Nations Convention to Combat Desertification
UNCED   United Nations Conference on Environment and Development
UNECA   United Nations Economic Commission for Africa
UNEP   United Nations Environment Program
UNESCO   United Nations Educational, Scientific and Cultural Organization
USDA   United States Department of Agriculture
WWF   World Wildlife Fund

Introduction

HUMANS ARE A PLANTING SPECIES. We plant trees in the millions and billions. On the face of it, humans plant trees for survival, to coexist and thrive on the planet. We plant trees to stimulate meaning, expression, and awareness, provoking poetry, art, and belief. We plant trees by necessity, to secure food, shelter, comfort, and fuel. But there is more to this. Humans also plant trees because we are very good at taking them down.

Afforestation is tree planting in otherwise treeless environments. Trees are woody plants, a form of classification. Treeless environments tend toward aridity, expressed as grasslands, prairies, and deserts—biomes collectively known as drylands, biomes replete with nonwoody forms of plant life. So the question this book asks is a simple one: Why do we plant trees in places without trees? And then: How do plants help us reckon with the environment? I ask these questions not because planting is exceptional but because it is so common.

Headline: "Ethiopia plants more than 350 million trees in 12 hours."[1]

The politics of plant life is located in the discrepancy between planting trees and removing trees because accurate tallies do not exist for global tree loss. Destruction carefully avoids precision, while tree planting poses no statistical or analytical ambiguity. It is simply easier to count tree units than to explain the quantity of destruction in a complex forest biome. Deforestation statistics only emerge from acreage and area, and find uncertain boundaries in logging claims and forest cover indicators in the margins of conservation reports. Deforestation creates slippery facts. On the other hand, afforestation endures as planting individual units disseminates calculation, and facts garner funds. Facts circulate headlines. Facts cannot be disputed. But facts without context are risky things.

Something comparably risky unfolds in the agreement that tree planting is always a universal "good." What ensues is an equally slippery agreement that numbers matter more than context, assuming a vacancy of life where projects unfold. Tree-planting campaigns assuage the moral and biotic difficulties posed by extant landscape conditions. As long as trees are being planted, trees can be counted. There is a simple method to investigate the advance of afforestation, setting up the goal of this book: study the largest and most impressive afforestation schemes across the world backward in time and outward in space.

What emerges is an accumulation of tree-planting initiatives that advance the same tactics across a century of environmentalism. The tactics entangle human and plant life, reshaping a set of linked but varied maneuvers. First, afforestation relies on tree planting as a form of earthly care that is free of criticism and full of approval. Relying on tree-planting "good" helps circumvent the most basic questions: Who is planting the tree, what species is specified, where is it being planted, and when? The second tactic weaves in dependencies, as miles of plastic dripline are disregarded as a mundane detail because any query of tree planting suggests criticism, a questioning of the cause itself. In times of crisis, this uneasy agreement dictates the third tactic, as projects pair with world politics, casting tree planting as a solution to varying crises from soil erosion and air pollution to extreme drought and more recently global climate change. Because no one is asking any questions, tree planting appears to be an environmental obligation or, more outrageously, a solution.

Tree planting does not moderate tree loss, although this is precisely the discrepancy occupied by global afforestation schemes. In a time of extreme loss, planting appears as a gain. Because it is unclear who gains, tree-planting programs keep humans safely on the outside of land-based crisis, charting an achievement that is only integral to some portion of society. The other share is persuaded by a mix of political and environmental authority. When a simplified version of plant life is arrayed across vast dryland territories, it reveals more about human relationships than it does about environmental ones. The clutter of tactics that exploits plant life as a respectable solution to anthropogenic change is rehearsed in watershed moments, which now include the great climate crisis.

Afforestation is an environmental venture that reveals an especially slippery slope because it represents progress in botany, forestry, and governance at the same time. The act of planting is not "environmental"; it is entirely anthropogenic. Planting is a radical rearrangement of the landscape regardless of extant conditions. It entangles humans by producing economic cultures reliant on models of standardization and governmentality. There is nothing ecological about digging a hole into the living soil where other plants are either established or lay dormant, just as there is nothing inherently ecological about inserting a woody plant into the land by will or force. Under the protection of technical know-how, no one asks who is planting the tree or where, let alone who might be profiting from cutting trees down. As long as trees are being planted, trees can be felled. Tree planting is further legitimatized through conservation tactics that warn of a tree deficit. In this way, anthropogenic tactics are inscribed into variable dune formations, introduced into the subtle velocities of airborne dust, and inserted into the rhizomatic horizons of drylands. The more conscious we are of the bias toward planting, the more chance we have of acting politically without sacrificing effective environmental action. Consequently, one of the other questions this book asks is why we value planting statistics without any indication of long-term success. Why are we planting trees instead of growing trees? The quick-fix mentality of planting units makes afforestation emblematic of our epoch. Afforestation is produced by a trajectory of understanding that pairs environmental action with environmental destruction, mingling hope and despair. In wanting to "fix" and "solve," we avoid the tough choices that force a deeper reckoning of ecology beyond economy, of being patient with earthly time, and with landscapes that are often hard on humans, especially drylands.

The tactics of afforestation pacify the aliveness of the dryland biome by prioritizing some plants over others. Plant life in drylands is iterant. Aliveness is marked and guided by cycles that accumulate as seasonal trends. Dormancy is the rule; roots spread underground persistently, while branches develop aboveground progressively. The difference between herbaceous (nonwoody) layers and ligneous (woody) strata are pronounced in drylands. The vivacity of each species is found in physical fragments, as seeds persist in the deep horizons of the soil, in layers of germination. Drylands transmit a different pattern of understanding between human and plant life that helps challenge statistical tree-planting campaigns, raising questions about other kinds of life and other ways of being human.

Throughout the book, I use the term *plant life* to remind readers that plants are living, breathing organisms. Deliberating on plant life does not lie in simply elevating plants to living status. It lies in the grading of human life as the *most* living species on the planet. It follows that human life is characterized through our unique capacity for narration, across centuries of divinity, linguistics, governance, and the arts. These particularly human advances are what help constitute our exceptionalism but tend to hold us back from other affiliations. For instance, neither the humanities nor the social sciences focus on plant life in order to examine society. Even archaeology is defined as the study of human activity and relies heavily on material evidence that backgrounds the living, breathing, behaving world. Plants endure as evidence or technique in design, just as agricultural exercises progress from strictly disciplinarian points of view or industrial prowess. What is left is a landscape free of the connections between people and plants, absent indigenous knowledge, ecological transformations, and other collaborative adaptations. What is left is very little.

We owe our atmosphere to plant life. The climate arises as plants condition the air, taking in carbon dioxide and making the oxygen we breathe. I am inspired by plant life because plants are the only organisms that connect the ground to the atmosphere, the soil to the climate. Ascribing agency to plants foregrounds their role in shaping our earth system, elevating plants to a force of nature rather than a background for human action, a stage for creature evolution, or a natural resource. In order to elevate plant life as a collaborative study between species, I describe plants through their formation—or activity—with an emphasis on behavior and communication. This aligns my study with the science of plant morphology, as opposed to plant anatomy, a discipline that relies on the aliveness of plants to engender scholarship. Aliveness shifts human assumptions away from objectification and the colonial record of plants as products of trade.

The behavior of plant life illuminates one of the most important misalignments of afforestation, what specialists call *species selection*. Careful decisions about which plants to include or exclude on recommended lists is the most important consideration of any planting project. Each list is inclined to use formal Latin binomials, organized alphabetically. Regrettably, species selection creates authority over local customs and extant plant life. It is also invisible to the muscle of governments, agencies, and other funding bodies. Species selection superimposes by design. This tendency is maintained by the authority of the sciences but brought into spatial practices without consideration. My suggestion is to suspend the usual order of increased productivity and one-dimensional knowledge systems that disregard plant behavior. Such authority suggests that some plants are more useful to humans than other plants that are thriving spontaneously. What is the value of environmentalism when it engenders negotiation, struggle, and indifference toward other species?

There is a passage in Aldo Leopold's journals where he expresses the importance of our relationship to the landscape through a lack of understanding for plant life.

The problem, then, is how to bring about a striving for harmony with land among a people many of whom have forgotten there is any such thing as land, among whom education and culture have become almost synonymous with landlessness.[2]

Leopold, like many who pay attention to plant life, describes landscape with care rather than precision, through relationships rather than numbers. Relationships are temporal; numbers are prescribed. Leopold sought the inclusion of plant life in order to foster his environmentalism, producing knowledge by experience rather than expertise. He praises plants for structuring the land and admires how the dryland prairie was built by "a hundred distinctive species of grasses, herbs and shrubs; by the prairie fungi, insects, and bacteria; by the prairie mammals and birds, all interlocked in one humming community of co-operations and competitions, one biota."[3] Leopold's capacity to notice the dryland prairie suggests a practice of patience; a slow appreciation of Paleozoic groundwater, rhizomatic deposits, ageless clones, and decisive seed; and the significance of dormancy in relation to the climate. The plants of dryland biomes have tremendous power at times, but such power is not always immediate.

The plants of afforestation, on the other hand, are powerless. They are not an agricultural necessity; nor are they are grown for consumption, timber, or wood pulp. Afforestation is tree planting without immediate economic benefit or long-term productivity. The statistics of each campaign legitimize afforestation as an environmental solution, masking its relationship to politics. Because the trend relies on tree-planting "good," projects garner support without actually having to stop violent forest decimation. We cannot offset deforestation of old-growth plants in humid biomes with unit-based planting in arid biomes. So the last question this book tries to answer is how tree planting in drylands became an indisputable form of environmentalism.

What afforestation reveals is that planting trees is sometimes just not enough. It is not enough to offset deforestation. It is not enough to solve carbon accumulation and rework the changing climate. It is also not enough to stabilize the current, disastrous moment in environmental politics. Planting a tree can be either one of the ultimate offerings to thriving on this planet or one of the most extreme perversions of human agency over it. We cannot just plant trees, and calculate units to escape, survive, or solve crises. This is a strategy of nonsense and repetition. Rather, we could safeguard extant trees in humid biomes; we could attend to fragmentation through reforestation, and insist that planting trees does not replace the slow accumulation of plant life. We could also learn to love drylands.

A Century of Afforestation

This book pays close attention to three cases that suggest the past, the present, and the future of afforestation. Chronologically, the cases begin in the early twentieth century, shadowing the American Great Plains; endure through the mid-century in western China; and are scheduled to blanket the African sub-Sahel from Dakar to Djibouti in the twenty-first century. Each afforestation campaign is united by its supracontinental range and aggregate in scale due to the authority of national and state authority. The cases share the same tactics but progress across different decades, registering a localized tree-planting project as a solution to a raging national or global environmental crisis, although the crises are often very far from the planting itself. The Dust Bowl years instigate American afforestation from Texas to the Canadian border, desertification prompts China's program, while the amalgam of food security issues encourages tree planting across the sub-Sahel. The reader might notice at this point that each context is vastly different in its temporal, biotic, and social contexts, which is only relevant insofar as, despite the differences, each landscape is classified as dryland.

Drylands emerge when aridity levels cross values of 0.5, instigating a change in the regional ecology due to a distinct lack of moisture. An aridity index is a measure of the ratio between annual precipitation and total annual potential evapotranspiration.[4] The force of low moisture thresholds is noticeable in the subtleties of plant life, as height is limited due to the twin effects of reduced rainfall and a general increase in temperature. Plants are lower and wider, rather than taller and denser, and tend to have little or no woody tissue, a character of herbaceousness. Woody plants that might otherwise grow to towering proportions under moist regimes become stunted and low in drylands. Along with the rise in aridity, soils become less and less fertile, which increases the likelihood of soil erosion. Dryland processes can be very slow by human standards, since each dynamic species in the environment is so dependent on minor thresholds of moisture. In short, dryland plants act as an indicator of planetary time, and thus planetary change. As an extreme example, imagine that the Sahara—one of the largest extant dryland biomes on the planet—was an ocean only ten thousand years ago. Such timescales are out of sync with human life but reiterate when climate changes begin to speed up. At a spatial scale, the area of terrestrial land occupied by the dryland biome exceeds 40 percent, an entanglement that reveals a dry planet getting drier, and consequently trending toward treelessness. But this does not mean these lands are infertile. Many drylands host annual and perennial growth, dormant seed, and spontaneous cover, which reveals that planting trees in treeless drylands is a kind of environmentalism out of whack with the nature of change inherent to the environment itself.

This is not to say that planting in dryland soils is always problematic, since a single tree cannot be equated to a continental plantation. There is a marked difference between planting a tree near your home or village in order to tend to it and direct its progress to benefit from shade, fruit, or resin; and appealing to international agencies for tree-planting funds in order to indiscriminately suppress and control land. What emerges from continentally scaled afforestation is not an ecological mandate but something deliberately territorial, meaning it inscribes political agendas on local and often marginalized populations. As a case in point, I came across more botanical than ecological literature in my studies of afforestation. Botany is the study of individual organisms, rarely including their interactions with humans and other species. Botany is the study of plant growth and development. Focus placed solely on the plant is unexpected in relation to forestry, especially forest building in the twenty-first century. The insistence on botany in afforestation is unexpected in current campaigns because projects land in very real places where other species persist. So, rather than express interest in the ecological connectivity of extant plant life, campaigns proffer lists of recommended species. Lists of individual, and often imported, species advance units. Copy-paste is not a smart environmental approach.

So why does afforestation persist? The answer to this question is found in a corrupt environmentalism that couples tree-planting "good" with a fraught reading of drylands. By definition, drylands are the context of afforestation, and attest to the scale of the "problem" because they occupy 41.3 percent of the total planetary land surface (see Table 1). It is simply too much to accept that such a large portion of the planet is less than hospitable to humans. Afforestation emerges as a "fix" that mingles with politics without debate, as projects advance the machinery of progress across unsustainable scales. Comparable with the alternative of exposing the evidence of a failed approach to drylands, maladministration, and corrupt authority, afforestation leans on tree-planting good in order to appear politically neutral, ecologically sound, and socially vibrant. It not only ticks a lot of boxes; it ticks them at the same time. Dryland scholar Diana K. Davis relates the suppression of drylands to sentiments of antinomadism.[5] I suggest that the suppression she articulates triumphs through afforestation. Moreover, it triumphs due to the suppression of relations that value plant life. Organized tree-planting projects and their exaggerated scope are proof that the common aim of afforestation is to pacify iterant cultures by replacing extant landscapes. It is essential to recognize that afforestation is not only prejudiced to certain plant species; it also blatantly prejudices the culture of millions of humans.

This book is written with a great deal of respect for plant life and a deep mistrust of afforestation. In trying to unpack why afforestation persists despite its environmental and social consequences, I probe the historical basis of organized tree-planting campaigns by examining tree-planting expertise, specifically the tactics of scientific forestry. The insistence on units emerges from the specialists in tree planting in ways that maintain tree felling. The tactics of specialist foresters rely on a *landless* culture. By contrast, I describe plant life through the lens of plant behavioral science. Plant behavior reframes the predicament of afforestation from a solely human journey to one that is more entangled, and more richly connected to the other species that structure landscape. Plant life counteracts *landlessness* because plant life is undeniably contextual. Behavior asks us to consider how plants grow and more specifically how they control what they do. How do plants know when to set down roots or send up a shoot? My approach is shaped by the conviction that studying plants makes us better at being human.

Table 1. The Dryland System

Dryland subtype	Aridity index	Share of global land area (%)	Share of global human population (%)
Hyperarid	<0.05	6.6	1.7
Arid	0.05–0.20	10.6	4.1
Semiarid	0.20–0.50	15.2	14.4
Dry subhumid	0.50–0.65	8.7	15.3
Total		41.3	35.5

Table redrawn after Uriel Safriel et al., "Dryland Systems," in *Ecosystems and Human Well-Being,* vol. 1, *Current State and Trends,* Millennium Ecosystem Assessment 1, ed. Rashid M. Hassan et al. (Washington, D.C.: Island Press, 2005), 627.

Our practices are set apart by how we acknowledge the spatial and temporal differences of other organisms, especially plants.

ARTIFACT, INDEX, AND TRACE

The three cases of afforestation in the sub-Sahel, the American prairies, and much of the Chinese desert and grassland steppe reiterate the question posed at the beginning of this chapter: Why do humans insist on planting so many trees in places that do not support tree life? In order to attempt an answer, the book is written in three sections: *Artifact, Index,* and *Trace.* Each section is divided into three chapters that describe the tactics that legitimize and advance afforestation in the particular context of the case under consideration. The final chapter in each section offers a description of plant life through the lens of plant behavior. Thus, the reader will not find a chronological account. Rather, the words *artifact, index,* and *trace* are used to describe common tactics regardless of sequence or context. Tactics that pile up and overlap deserve to be pulled apart because they reveal that afforestation is not ecological; it is deliberately political and distressingly social.

The first procedure I describe is *artifact,* a framework that sheds light on how the study of plants generally condenses aliveness. The word *artifact*—an object made by a human being—suggests that humankind considers plants a static product rather than a living organism. *Artifact* is a term that encompasses how plant life is forced awkwardly into human worlds and aggressively into biotic ones. Specifically, the creation of a predictable artifact engendered specialization of select plant parts: the berries, seeds, bark, or tonic that create resources. I extrapolate that the insistence on plant parts gives way to the insistence on counting units. Rather than reiterate the history of colonial exploitation, I suggest a lineage of botanical theory and plant morphology that respects the aliveness of plants. When *living* plants are studied, context simply cannot be removed or abstracted. Here, scholarship survives as a series of deeply committed connections between human and plant life. The lineage is only one of many but is offered as a means to reassess our botanic inheritance and appreciate counternarratives. What is remarkable is only that studying living plants is a marginal history, so I pursued the writing with the hope and conviction that the sciences built of speculation over millennia can be renewed and benefit from sharing exploratory, rather than exploitative, encounters.

The second procedure reveals the advance of tree units. *Index* is a word I use to register the economy of trade and accumulation tallied by the mechanics of lists and manuals. The word *index*—to point out and make known—suggests a method of both identification and standardization. The formula of trading in units engages the familiar protocols of designers, who tend to shrink at the authority (and impenetrability) of scientific discourse and promote resolutions based on previously established and often misleading metrics. It is precisely the translation between trust in science and faith in problem solving that is at the heart of all formal aesthetic practices, and of the design professions in particular. Growing trees demands human stewardship, someone who cares for the plant and gives it additional nutrients, water, and protection in order for it to survive the radical insertion into foreign soils. Thus framed, I conceive of afforestation as a kind of horticulture, the practice of plant cultivation and management, related to gardening. I include the forestry professional as a designer because their work is inherently biophysical, a feature of continental horticulture. The aggregate effect describes how the value of plant life is arrested as an object of study in *artifact* and a unit of profit in *index*. One practice exploits the plant itself, the other, knowledge of the plant.

Trace is the third procedure, marking the transition from abstraction to the material, physical landscape. The word *trace*—to give an outline of—suggests how tree planting materializes as a means to gain jurisdiction over territory, setting up a response to crisis that appears real and necessary. Spatially, the question lies in the disparity between forest plantations and extant drylands. What *trace* reveals is that the brutality of afforestation is found in the failure to create lasting relationships. This is the difference between planting trees and growing trees. Instead, afforestation is embossed on drylands, leaving physical marks and scars that appear in real places, plants left to thrive or wilt by design. Whether we are talking about tree planting or tree growing, the model of supracontinental afforestation operates on the radical assumption that horticultural techniques can be applied at ecological scales. In this final section, the creation of the botanical *artifact* expedites the professional distribution of *index,* as units guarantee the rejection of context through *trace.* Afforestation is established in this maneuver, an environmental construct so pervasive that it can be rehearsed anytime, anywhere.

Throughout the book, plants emerge as living, breathing, earthly interfaces between the soil and the climate, between sociocultural and biophysical conditions. Therefore, the plant is the designer of success and failure in the drive to afforest, reforest, and deforest. Plant life extends beyond noticeable boundaries and physical sites, and includes the temporal, buried, and symbiotic sites that remain unexpressed and unaccounted for within procedural structures and political determinants. Because species live and thrive through human abstraction only means that there is a lot of life on the planet that is neither fragile nor in need of our protection. Life collaborates by tinkering with tactics and reversing political protocols like afforestation.

The underlying tactics of *artifact, index,* and *trace* endure in each case, and are only isolated in each section to shed light on the consistency of the procedures. Afforestation is common to all dryland systems, rendered as a common solution to climate hazards that range from drought to flood (see Figure 1).[6] The simplifications of ecological difference are overcome by design. In order to show the unity of approaches, I paired one tactic of afforestation with one case.

But these tactics are not isolated; the reader will notice that *artifact, index,* and *trace* are common to all three cases. I do this to overcome the assumption that afforestation is an outdated or obsolete agenda. On the contrary, the same ecological manipulations continue to press into the twenty-first century, and find their way into climate change policy. For instance, the first section considers the African "Great Green Wall" (2005–present) and pairs with *artifact,* suggesting that the exploitation of plant life is not a bygone story of colonial abuse but a persistent thread in extant biopolitics. In the first chapter, the reader will notice that the impact of studying plants in parts gave rise to trading plant products, a procedure that continues to distance plant and human life. I begin with this case because the formula reappears in twenty-first-century environmental mandates, as tree units advance the campaign to afforest African

Continent	Country	Desert	Hazard
Asia	China/Mongolia	Gobi	Drought, dzud, flood, landslide
	China	Taklamakan	Sand movement, landslide
	Central Asia	Kara-Kyzyl Kum	Dzud, drought, landslide
	Iran	Lut	Drought, flood, dust storm
	India/Pakistan	Thar	Flood, drought, locust
Middle East	Arabian Peninsula	Arabia	Drought, locust, flood
	Syria, Jordan, Iraq	Syrian	Drought, flood
Africa	Africa	Sahara	Drought, dust storm, locust
	Southern Africa	Kalahari/Namib/Karoo	Drought, flood, sand movement, quelea
Australia	Australia		Drought, flood, wildfire
North America	Mexico / United States	Chihuahua, Mojave, Sonora	Drought, flood, wildfire, coccidioidomycosis
	United States	Great Basin	Flood, drought
South America	Peru, Chile	Atacama	Drought, flood, landslide
	Argentina/Chile	Patagonia/Monte	Flood, landslide
	Brazil	Noreste	Drought, coccidioidomycosis

FIGURE 1. Summary of climate hazards in major drylands. Disturbance is a natural occurrence across drylands, ranging in effect from flood to drought. Plant life in drylands tends to be well adapted to contextually specific hazards. Figure redrawn after N. J. Middleton and T. Sternberg, "Climate Hazards in Drylands: A Review," *Earth-Science Reviews* 126 (November 2013): 48–57.

drylands. The second section describes the "Prairie States Forestry Project" (1934–42) through the *index* of professional and expert forestry.[7] At the turn of the century, federal policy mobilized tree planting, a political subversion disguised as a conservation protocol. Afforestation endures by mandate, within the rigors of scientific tradition. Finally, the third section explores the largest acts of horticulture across the broadest timescales. *Trace* is a procedure that frames the "Three Norths Shelter System" (1978–2050) because the project unsettles spatially and temporally and demonstrates the endurance of planting in otherwise treeless biomes.[8] In the Chinese context, the margins of specialist forestry expand to include afforestation, encouraged by the authority of global aid models that count units to proliferate headlines. Each decade of afforestation in the Three-North region reiterates the procedures of *artifact* and *index*. The lasting cultural and biophysical effects hide in ways that deserve notice.

The persistence of afforestation is found in the progression from *artifact* of trade to *index* of expertise, and emerges as an environmental *trace* because human life is dislocated from plant life. The public perception of plants as a measure of human prediction and calculation renders passivity that serves first as the basis for perceiving the world and second as a means to control it. As designers, we inherit the exploitative model of botanical speculation, tied to the regulatory structures that encode afforestation. Cast in this light, landscape practices are put into sharp focus, as current traditions gain traction from a history of authoritative exercises that position the human outside "design." The foundations of afforestation help enlarge our understanding of how deeply the expert human—scientist, forester, or designer—is entwined in the process of extracting or suppressing the aliveness of plant life.

The drive toward afforestation highlights the divide between shared knowledge with other behavioral organisms and the perversion of using careful insight for crude resource exploitation. As a result, increasing specialization transfers any agency on the part of plants and relocates it to humans. To counteract this, I pay close attention to particular plants of afforestation, to help untangle the notion that plants are immobile units of expansion. An inquiry into what we know about tree planting must be met with a study of what we do not know about plant life. The long, entrenched scientific and regulatory arrangements that pacify plant life are examined in each section to highlight the difference between how the plant is known and how plant life makes itself known.

WHAT IS PLANT LIFE?

The word *plant* emerges from two Latin origins: *planta*, to sprout; and *plantare*, to fix in place. One origin insinuates movement, the other stasis. In French, *plante* offers the same inconsistency, so that the word itself is caught between its meaning as a noun that indicates development and a verb that implies permanence. This modest etymology suggests an entry into our relationship to planting, as plants are objectified as a fixed form of knowledge, such that their aliveness is no longer a subject worthy of human speculation. The plant is wedged between worlds: aggressively weedy or strikingly ornamental, beneficial or useless, either despised or desired as it is naturalized, pacified, or capitalized. All ensuing engagement with plants necessarily chronicles the project of domestication whereby the human subject cannot resist being the active agent, endowed with power and armed with prediction and, more recently, as a pioneer of innovation. Each step—each accumulation of expertise—transforms plant life into a measure of human knowledge. As *plantare* achievements continue to dismember the organism as a whole, plants are recast as tools of science. Images of the plant-object are fixed in the human imagination, a world of forms.

While the objectification of "nature" has been suggestively undone by science studies, the plant-as-object still finds its way into spatial practices such as geography, forestry, and landscape architecture. But this will be explained further. For now, it is important to note that the ensuing results lead to environmental projects that treat plants as tools, specifications, and statistics. The act of planting materializes plant life absent agency, movement, behavior, or fundamental biological activity. Rather than the fixity of *plantare,* consider instead the definition of *planta*—to sprout. This verb status fittingly attends to plants by means of their activity and behavior, a mode of being that helps shed the perception of fixity. *Planta* reveals that a plant is a process, a swarm of activity, and a dynamic planetary force. In light of the cascade of concerns that mark our time, recovering the relationship between plant and human life is more important than ever. I am suggesting that animating terms allows us entry into a more inclusive definition of plant life.

There is significant debate over the definition of *plant,* the most common term used to describe green, photosynthesizing organisms, including some fungi and algae. Biologically, plants are referred to as "eukaryotic photoautotrophs," while the *Oxford English Dictionary* gives the following definition: "A living organism of the kind exemplified by trees, shrubs, herbs, grasses, ferns, and mosses, typically growing in a permanent site, absorbing water and inorganic substances through its roots, and synthesizing nutrients in its leaves by photosynthesis using the green pigment chlorophyll."[9] Can one definition do justice to the almost 400,000 plant species known to science? The sheer number of plants in the kingdom *Plantae* entirely overwhelms the imagination. Defining "plants" gives rise to debate and erroneousness, such as the reference to growing in a permanent site, a feature of the standard definition that disrespects the radical behavior of plant life: the migrations, invasions, and movements that animate our relationship.

Advances in plant behavioral sciences offer an entry into the claim that humans are a planting species because plant life is appreciated and studied as an organism rather than as a subject to be "broken down" or "sorted out." Including behavior reminds us that plants cannot exist without context: the encoding, storing, and recalling of information by the plant through the environment, the incessant interactions with the material and nonmaterial world. Some scholars reference plant evolutionary adaptation as a form of "remembering," encouragingly disrupting questions of permanence or stasis.[10] Scholarship motivated by a closer reading of plants as behavioral organisms reminds us that science is an exploration, an opportunity to find deeper meaning founded on a desire for mutual understanding. It is also a reminder that deeper meaning includes environmental variability because plants link the climate to the soil.

> To contend with environmental variability, plants often show considerable plasticity in their developmental and physiological behaviors. Some of their apparent choices include: when and where to forage for nutrients and where to allocate those nutrients and derived organic molecules within the organism; when and what organs to generate or senesce; when to reproduce and the number of progeny to create; how to mount a defense against attack and in what tissues or organs; and when and where to transmit chemical signals to surrounding organisms.[11]

This leading paragraph from a controversial article from *Trends in Plant Science* (2006) was written as a proposal to shape the emerging field of plant neurobiology. The article brought together a group of scientists who argued for a discipline that could combine the efforts of diverse plant sciences in order to answer a common question: How do plants acquire information from their environment, both abiotic and biotic, and integrate this information into responsive behavior? The implication that plants make choices, mount a defense, or communicate with other organisms calls to mind pseudoscientific claims, and the particularly garish label of "soft" science.[12] How could *neuro-* be applied to *-biology* in organisms absent neurons? More than twenty plant biologists responded with a scathing critique of plant neurobiology for its "superficial analogies and questionable extrapolations," cautioning that there is no clear benefit to the scientific research community.[13] For scientists with brains, the suggestion was not only outrageous; it sparked wide controversy and animated rejections.

The slippery margins of fact are revealed in these contributions; the question of defining plant life shifts the contours of science by determining that plants behave. Terms such as *plant signaling* and *plant intelligence* are sneaking into academic and scientific journals, as university departments and laboratory societies begin to accept the provocation that plants might be much more attuned to the world than any claims previously suggested. The authors render plant life beyond a commodity or a resource, and well beyond the desiccated sample, by stimulating environmental variability, what I refer to as context. Context reiterates in this book to appreciate the difference between sites, the difference between looking at a landscape and being very much a part of it.

The response of the authors in relation to context acknowledges that the aliveness of plants is but one step toward refining collaborative earthly relationships.

> We are less concerned with names than with the phenomena that have been overlooked in plant science, which, in our opinion, need to be addressed to truly understand plant operation, particularly in an era of outstanding new technologies.[14]

The "overlooked" phenomena are the behaviors, signals, and exchanges that plants integrate in order to sense, grow, and adapt to their environment. Thus, the term *behavior* is not inevitably related to humans and other creatures; rather, the proponents of plant behavior explain decisions through chemical vocabulary that includes the ability to sense humidity, light, pressure, and volume at the same time as nitrogen or oxygen, salt or phosphorous. Plant behavior is the movement of a root tip around a stone, the internal reallocation of resources to external stimuli, and the agility to remain sessile while defending and feeding. The authors ask us to consider that neurons are not a requirement for learning, which raises the question of how plants sense change in their environment and respond by altering their development. Attention paid to plant life helps us learn to value evidence that is hard to objectify: the systems that encroach on commodification. The facts that resist need our attention more than ever.

One of the most resonant "facts" that resists our attention is concealed under our feet, in the significance of plant roots in context, the root-soil interface that is often called the rhizosphere. Therefore, one of the ways I expose the myth of afforestation is to spend time describing behavior in the rhizosphere. Plant life advances underground and helps deviate from the naive insistence that plants are only worthy of their aboveground attributes. While I was aware of this before writing this book, attention to the rhizosphere established a method, a means to study each plant without culling through the vast literature on aboveground "parts" alone. The inclination toward roots was amplified by plant behavioral studies. According to Stefano Mancuso, one of the coauthors of the aforementioned study, the most excitable cells are found just behind the root tip. This is a hypothesis first proposed by Charles Darwin in *The Power of Movement in Plants* (1880), a treatise that asserts plants can sense their position in the world.[15] Mancuso builds upon the evidence, arguing that plants sense their environment stimulated by behavior *below* ground. Mancuso argues that plant behavior validates a unifying mechanism across all living systems that enables the processing of information and learning. What is most contentious to the argument is that because plants can process and learn, it could be that intelligence does not require a brain.

The incredible disparity between aboveground and belowground research is found not only in the language of sentience but in the view that plants might actually be more perceptive than ever imagined. Unlike antiquated ecological concepts of pure competition in nature, behavior escalates the capacity for collaboration because plant life does not differentiate between organisms. Both a nematode discovered deep in the soil by a root tip and a human digging into the surface of the soil stimulate conscious activity in the soil.

> Although not seen, the growing roots system is in a remarkably dynamic state. Sensing mechanisms operate that enable accurate navigation of the soil mosaic. Soil organisms, pockets of water and minerals, soil characteristics and crumb structure, pockets of gases, soil chemical, and stones are all recognized, and action taken to exploit or avoid.[16]

Attention paid to the vivacity found in soils broadens our understanding of life on earth. Once the plant is understood as a behavioral organism rather than a tool, we become entangled in the specificity of behavior.

Consider the range of scholarship undertaken by the coauthors of the original paper: Eric D. Brenner's lab studies lateral root formation, Rainer Stahlberg analyzes uptake in roots in order to address long-distance signaling, Elizabeth Van Volkenburgh detects root growth and function through what she terms "whole-plant behaviors," and, in particular, František Baluška publishes prolifically on root organization.[17] Each scientist explores the mechanisms driving plant movement and describes behavior through the intelligence of the root system. Richard Karban's *Plant Sensing and Communication* (2016) is particularly valuable to this lineage. Karban considers sensing a clue to behavior, and offers examples that range from the effects of touching on growth to rhizomatic volatile signals and eavesdropping leaves. An entire chapter is dedicated to communication with microbes, with particular focus on underground relations, whereupon he states: "Far from being inanimate, plants sense, communicate, and show sophisticated conditional behaviors that we are just starting to appreciate."[18] Upon closer examination, each scientist is exploring the world belowground in order to further their research, tracing signaling between plants and other organisms to the depths of the soil where roots and rhizomes progress. In turn, I am inspired by the suggestion that plant life does its living in the rhizosphere.

Plant behavior extends our appreciation beyond the individual plant and begins to define plant aliveness within a broad association of species, influences, and forces. Behavior is never isolated, exposing the violence of counting individual tree units. Research offered by ecologist Suzanne W. Simard is especially valuable in this regard, since her examinations distinguish behavior at ecosystem scales, by paying attention to signals between living plants. Simard ties the submicroscopic world of atoms, molecules, and mycorrhizal networks to the macroscopic world of droughts and erosion.[19] I prefer to imagine Simard's work as an exploration into the rhizosphere as a rich zone of exchange, where mycorrhizal fungi link the roots of different plant hosts, forming associations that are just coming to light. Simard breaks the plant apart, not for specialized resources but as a collection of slowly dividing cells: "Today, plants are commonly recognized as microbiomes—where villages of collaborative microbes live in and on their roots, stems, and leaves forming interaction networks."[20] As an explorer, Simard suggests that plants, when defined as microbiomes, challenge global systems of exploit, a curiosity reminiscent of Lynn Margulis's scholarship on collaborative integration: "Life's nature is to interact with the material world to incessantly integrate its components, rejecting, sorting, and discriminating among potential food, waste or energy sources in ways that maintain the integrity of the organism."[21] Using accessible language, and armed with the tools of empirical study, Simard and Margulis are part of a lineage of scientists engaging with the living world outside obvious extractive resource potential. Instead, a common future between species slowly emerges where singular, short-term solutions might appear to be perverse when the living environment is considered for how it holds us together. These are some of the scientists, explorers, and instigators that I reference, in order to help refine an active definition of plant life.

It is my assertion that paying close attention to how plants are understood or misunderstood always lands us in the rhizosphere. So much of plant life is concealed below the surface, offering an opportunity to refine our environmental aspirations. Paying attention helps disentangle institutional abstractions, while attention to the rhizosphere sheds light on the fallacy of afforestation, and our urges as a planting species. Is the life of the plant sealed within its own system, or does it expand outward, merging with the environment in association with other plants, animals, and humans? Where do roots end and the soils begin? While seemingly ordinary, the aggregate difference between planting a tree with a bound root system and facilitating seedling development with thousands of rootlet strands begins to answer the question. It also engenders a more equitable explanation to the question: What is plant life?

If afforestation projects accumulate evidence as a means to captivate global attention, then it is especially important to refine and imagine what afforestation initiatives *leave out*. This is why each section of this book explores the behavior of plants and the study of roots and rhizomes as a means to cut across scientific literature and prioritize or reflect on how the concealed formation of plant life is included or excluded. Through the lens of plant behavioral sciences, afforestation is cast as a remnant product of the Holocene, a clever political tactic of the times, historicized through the linear trajectory of administrators rushing to identify a problem in order to fund specialists who can provide answers. Afforestation is so common that it helps produce an understanding of how we compose the worlds we inhabit, and acknowledges that these questions are not only political but extend the limits of what usually counts as politics.

ARTIFACT

The Great Green Wall of Africa imagined by the procedures of artifact, index, and trace.
Image by the author.

Artifact suggests a reading of plant life arrested by botanic authority and industrial standards. The plant-as-artifact is naturalized into human worlds through procedures that trade, study, and evaluate the plant-in-parts, extracting aliveness from plant life. This section requires a historiography in order to provide insight on these procedures because extraction lies in a thick lineage of colonial abuses that endures in the loss of understanding between human and plant life. Afforestation is an extreme version of our loss, as woody plants are valued over those thick iterant mats of perennials, frutescent thickets, or annual grasses and forbs that advance in drylands without any immediate human advantage.

Through the lens of *artifact,* plants are rendered productive or unproductive, and emerge as property or resource through a series of procedures: First, the colonial economy in the seventeenth and eighteenth centuries extracts plants for their useful parts, replacing traditional, land-based knowledge with extractive and exploitative maneuvers.[1] Next, trade in plant parts engenders specialization, subsidizing taxonomic reform and authorizing the global exchange of plant-based products. As a result, the spices, medicinal drugs, perfumes, or dyestuffs of commerce chart social progress rather than spread plant knowledge, reducing plant life to a metric of human progress. This proliferation of static knowledge—the artifact—positions planting policy in the twentieth century, as the plant-as-artifact makes its way back into the physical, material world. The problem is that the plant only makes its way back into human worlds as a product, not an organism. Each step in the process of objectification yields single-species, single-commodity redundancies, often expressed as grid-and-row plantations. These tactics reiterate in twenty-first-century environmentalism through design professions including forestry, environmental planning, conservation efforts, and domestic practices. The radical consequences of afforestation are a template of our loss.

Artifact alludes to the outcome of increased specialization that dismembers plant life in order to make the unruly plant world more legible. For instance, binomial classification engenders a management style that reduces plant behavior and encourages distance from the living plant. It also significantly underwrites global expansion policy from colonialism to environmentalism by replacing indigenous knowledge.[2] Within this tradition, plants are generally studied as to their kind, their structure, or their economic value, casting aside aliveness, as the plant is reduced to recognizable parts: wood, seed, resin, bud, flower, fruit, and nut. As such, planting design is also illuminated for accepting this tendency to emphasize plants as a resource rather than an organism. In order to appreciate the fuller origins of botanical thought in this context, it is necessary to include the creative endeavors of plant morphologists, a science that acknowledges plant life outside pure utility and product making.

This section is drawn into alignment with select treatises that advance a consideration of plants as living organisms and earthly collaborators, in order to overcome the desiccated sample, the plantation, or the ecosystem service. In comparison to popular botanical accounts of plant acquisition, introduction, or classification, I account for less exploitable studies that run counter to the plant-as-artifact. Tentative collaborations in botanical studies can be neither generalized for replication nor simply appropriated to create standards. As a result, the specificity of encounters between human and plant life is relevant to recovering familiarity with plants and informs how we might work with plant life in the future. In particular, a reconsideration of early botanical treatises and the theory of morphology are positioned as an opportunity to expand our knowledge of plant life beyond the more durable histories of exploitative botany. The counternarrative embraces aliveness and meets the challenge with an emphasis on personal engagement and firsthand experience.

In the African Sahel and sub-Sahel, the inheritance of *artifact* inscribes peripheral agendas into local politics, as control accumulates through the persistence of tree units. Livelihoods do not depend on the desiccated, classified plant-as-artifact. Rather, human and plant relations across dynamic dryland biomes depend on a careful measure of moisture, stewardship, and care for the few woody plants critical to secure fuel and food. The power of Africa's Great Green Wall lies in our botanic inheritance because colonial imposition and extraction not only manipulated the historic relationship between plant and human life in the past; it still does.

1

The Problem of Parts

In considering the distinctive characters of plants and their nature, generally one must take into account their parts, their qualities, the ways in which their life originates, and the course which it follows in each case: (conduct and activities we do not find in them, as we do in animals). Now the differences in the way in which their life originates, in their qualities and in their life-history are comparatively easy to observe and are simple, while those shewn in their "parts" present more complexity. Indeed, it has not even been satisfactorily determined what ought and what ought not to be called "parts," and some difficulty is involved in making the distinction.

—Theophrastus, *Enquiry into Plants*

IN THE FIRST LINES OF *Historia Plantarum* or *Enquiry into Plants* (350 B.C.E.), Theophrastus (ca. 370–ca. 285 B.C.E.) explores the difficulty of segmenting the growing plant into parts. Because the problem of "parts" was the primary motivation for writing *Enquiry*, it helps unpack the tension created by the procedures of artifact. Theophrastus studied plants to gain understanding, offering an alternative to studying plants solely for consumption or use. *Enquiry into Plants* suggests a moment at the edge of standardized scientific inquiry, a supple and less segmented intellectual basis from which to consider different forms of knowledge production. The "problem of the part" is resolved in *Enquiry* by two methods that will be further elucidated: by identifying distinct periods of permanence and impermanence, a behavior embedded in the term *orme*; and by prefiguring ecology and conceiving of context, or the link between plants, the environment, and other species. Theophrastus was perplexed by "parts" of the plant because it was unclear if some parts, such as the bark, actually belonged to the plant or not. He speculated on the plants' ability to release some parts and retain others. These breaks in continuity arise when studying organic formation, a botanical hitch that resonates to this day: can plants be studied absent their aliveness?

The pre-evidential status of this speculation is entrenched in Stoic philosophy, a reading of *Enquiry into Plants* that engages other kinds of aliveness. The philosophy of Stoicism asserts that behavior constitutes life, a mark or indication that what is real has being.[1] Attention to behavior is supported by the peculiar tendency of Theophrastus to describe plants using the ancient Greek *orme,* meaning both reasoned choice and irrational impulse.[2] *Orme* is a term typically used in reference to humans, revealing the importance of plant life to Theophrastus. Accordingly, if plants are real, then they have agency. The oldest known botanic treatise resolves the question of plant parts by paying attention to development instead.

Consider the difference between plants that arise spontaneously and those that are planted deliberately. A cultivated plant requires human agency and often requires support to survive. According to Theophrastus, wild plants grow with their own *orme,* a decisive individual behavior. Plants exert an effect on the environment, which describes an ecological position.[3] By differentiating between cultivated and wild plants, different sections of *Enquiry* articulate the tendency of the plant (which had its own purpose), the tendency of the environment (which may act with or against the plant), and the tendency of the human, which usually achieves a goal that is inconsistent with the plant and the environment.

Entire books in Theophrastus's oeuvre are devoted to the comparison between plants that grow with the help of humans and those that prosper without, including *Book II: Of Propagation, Especially Trees* and *Book III: Of Wild Trees.* Both treatises represent the earliest attempts at classification, distinguishing between plants that require deliberate planting and those that are spontaneous, noting a disparity in the way in which their life originates.[4] The role of dispersal between species is explored by observation, describing a set of associations outside human care in part because development is hinged on a plant having its own agency: "Any plant may be either wild or cultivated accordingly as it receives or does not receive attention."[5] This is a specific kind of advocacy because it promotes a certain independence that plant life no longer enjoys. *Enquiry* differentiates between plants that advance with human attention and those that do not, but in both cases, he maintains that the plant is always developing with environmental inputs, a fundamental tenet of ecology.[6] As a result of the differentiation, Theophrastus not only set the stage for ecology; he cautions the excluding and exacting influence of humankind, and presages the impact of human agency on the natural world.

In order to resolve the "problem of the part," *Enquiry* links the plant and the climate. While this extends the association with ecology, it also challenges the distinction of cycles, or periods that require permanence and impermanence. Plant parts are attendant to changes in the weather. To overcome the part in relation to external influences, the treatise engenders a distinct vocabulary found in the development *between* parts, dismissing any similarity between shedding fur and feathers, since the plant "has the power of growth in all its parts, inasmuch as it has life in all its parts."[7] By registering an active presence in the plant for the first time, the entire plant emerges against more familiar, divided forms that segment roots from fruits, nuts from flowers, resin from bark, and so on. The understanding between human and plant time is associated with growth, articulated by ongoing development. Unlike humankind, plant life evades knowledge in parts through continual growth and rejuvenation. The emphasis on growth in the entire living plant extends underground, although *Enquiry into Plants* seems to struggle with the term *root,* which suggests a parts mentality. Mentioning species as varied as *Olea* sp. and *Scillia* sp., the prose even speculates on a form of classification that could be based instead on root types.

Botanical thought is plagued with comparison to zoology. This includes our propensity to compare human characteristics with plant development, in such axioms as *all oak and iron bound,* or to anthropomorphize plant performance, like describing a weeping plant as sad. In *Enquiry into Plants,* Theophrastus extracts much of Aristotle's biological terminology and applies it to plant scholarship, including flesh (plant pulp), backbone (leaf rib), and heart (pith wood), forever assimilating plant and animal and human kinds. The work of philosophy into the natural world is substantiated by the repeated use of Theophrastus's term *relationships,* which is derived from Aristotle's *Historia animalium* (*History of Animals*). The title, of course, also attests to Theophrastus's ambition to admit his teacher's work into the plant world. Even the terms *behavior* and *movement* evolve from Aristotle's discourse of animal observation. Yet Theophrastus was not interested in a deeper understanding of the distinction between animals and plants like his mentor Aristotle, or in trying to force a conception of the plant into a zoological framework.[8] It seems as though the use of terms was only a form of efficiency. Accordingly, he claimed that making comparisons was a waste of time, and he seems impatient for effective appreciation of plant life, with fewer obstacles.

By using the term *originates,* Theophrastus activates his thesis, referencing the cause of biotic emergence by describing the first stages of the plant as a seed with reproductive force. This is *orme.* Plants are as significant as animals but distinct due to the acknowledgment of development rather than the isolation or identification of parts.[9] Therefore, the challenge of *Enquiry* lay in the articulation of transition, the development between segments or stages— from seed to sprout, to shoot—in which he identified the critical role of the root stem. Transitions offered a unique challenge that was not categorical to humans and other creatures, since each measure is reliably shed and renewed in plant life.

Theophrastus praises plants that endure regardless of our care, suggesting that a differentiation between spontaneous and cultivated habits is a distinction worthy of further consideration because it either includes or excludes humans. Such a distinction lays the foundation for a more inclusive approach to plant life, as *Enquiry* asserts that wild or spontaneous plants are stronger than cultivated or planted ones by citing resistance to cold, endurance to weather extremes, and fruit-bearing capacity.[10] In fragments, he laments why the fruit of wild species is unpalatable but prolific. Through his treatises, Theophrastus aligns spontaneity with "Nature," but he appropriately frames the investigation as a question rather than a response, an inquiry that continues to this day.

> Concerning this very matter there might well arise the general question as to whether we should regard as the natural type those plants that grow of themselves or those that grow under cultivation, and in which of the two lies the development that is according to nature. This is the same as, or rather a part of, that other question: whether Nature has in herself the vital principles. We speak also of one thing as being natural which comes into being of itself, while we call something else artificial because it arises from some impulse outside itself.[11]

The effort to define *natural* is not unlike Theophrastus's other efforts to reason, which are marked by the ambition to distinguish plant life within the natural sciences. At the time, the status of plants was characterized by *Scala naturae* or Aristotle's Ladder, a hierarchy that placed plant organisms underneath insects, sponges, and mollusks, only one rank higher than inanimate matter. Aristotle's suggestion reveals with confidence that plants were hardly living organisms. By contrast, Theophrastus broke with static notions of plants by paying close attention to the growth between parts, or behavior, regardless of whether the behavior is welcome in the human terms. The *impulse outside itself* occurs without external cause, related to spontaneity, or *sua sponte,* which translates to "at its own prompting" in Latin.

In Theophrastus's words, the force of things is more than a sophisticated posthumanist theory; it challenges technical categories and classifying language. The culminating effect is that many plants grow entirely of their own accord and do not rely on human agendas. Attention shifts to plant life as an intermediary within the environment. Yet agency is rarely given over to the plant because any such suggestion diminishes human agency, so the power of movement is allocated back to the human. This swapping of positions is assembled by design as human intervention and protection are elevated above plant behavior. The problem of parts is manifold, as it reiterates in ongoing debates of conservation, whereby plants that do not require human attention fall outside the botanical fold and tend to be relegated to aggressive or invasive status. As a result, the plants that enter into our lives are only those that are deemed productive to the human project. This debate will be addressed in the next section; for now it is worth appreciating that Theophrastus admired plants through their aliveness, absent human judgments or the abridged binary of productive and unproductive plants.

The premodern, nonmodern, unmodern categories of *Enquiry* are not the celestial observations of a dreamer; nor are they speculations of pseudoscience. They put to task the very authority of modern scientific pretension. It is perplexing that a scholar as prolific as Theophrastus, who turned his attention to so many topics that are central to the environment—from weather signs, sweat, odors, and stones—should nevertheless continue to be so rarely referenced in climate change literature or considerations of the Anthropocene. If plant life can be registered as an earthly presence, it is worthy of consideration beyond use, medicinal value, and medieval ritual. While *Enquiry* lingers as a part of botanical history, I consider Theophrastus's treatises an unexplored contribution that lays the foundation for an expanded consideration between human and plant life.

Theophrastus helps discover a context for plant life that is not driven by problem solving or regulated education systems according to institutional financial support or national innovation systems. The example of *the space between parts* underscores the importance of how parts fragment knowledge, a recognition that should not be conflated with progress. The key challenge of our time is whether plant knowledge can be freed from the production of artifacts and regain its aliveness.

SCIENTIZING PLANT LIFE

The procedures of artifact provoke value in parts. The plant is arrested as a measure of human endeavors as cultivation continues along a scientizing sequence that values trade above aliveness. Because aliveness does not matter to an economy of parts, the plant is absorbed into the world through plant samples, extracted remnants, and binomial epithet. The scientization of the plant is found in the fact that plants gain economic value outside context, and supports the knowledge economy absent any earthly connection. The problem of parts is overcome by studying plant samples instead of living organisms. Consider that a theory of spontaneity must necessarily emerge by studying living plants. In the hands of Theophrastus, plant life is varied and active. It supersedes human agency and reaches into the environment to gain space, secure nutrition, and advance its own kind. On the other hand, plant samples, parts, and pressings require extreme precaution to preserve the desiccated specimen, casting plants as fragile, weak, and requiring human protection. The artifact arises for scientific verification.

Early botanical transmissions were codified through samples and pressings, offering a reduced reading of plant life. As a result, plants were no longer an inquiry or a philosophy but an encounter with objectification. Here, the only knowledge being transmitted was of a fragmentary nature, designed to reveal value in ideal form or economic benefit. Why establish a sample or conserve a pressing if the plant had no immediate productive value for humans? Therefore, the process of transferring plants across diverse ecological and social situations confirmed research in cultivated plants, necessitating an excess of institutions to confirm or reject imported luxuries or cash crops such as maize seed, nutmeg saplings, or Peruvian bark.[12] At the same time, the plants that proliferated under the cracks of each great empire were rarely, if ever, codified, identified, and counted, activating the association with useless or nuisance plants such as those we now call weeds.

Casting plants as fragile material in need of human care may be characterized as a general condition of early science, the residues of which persist today. Unlike courageous plant explorers, or indigenous customs, taxonomists pursued a neat, careful scholarship that preserved the crumbly herbaria deposit in cabinets, drawers, and bound volumes, whereby any plant can be pulled out and produced without delay.[13] Tremendous care must be taken with plant samples, necessitating white gloves in darkened, moisture-free environments, accessible only to those trained in the handling of such artifacts. Systematic botany established a common language for the natural sciences through binomial nomenclature, which facilitated prediction of the unruly plant world. Binomial nomenclature was developed by Carl Linnaeus, also known after his ennoblement as Carl von Linné, who did not invent the system altogether but borrowed and refined the ideas of others, creating a standard system for naming organisms composed of two terms, the first indicating the genus and the second the specific epithet. For instance: *Homo sapiens* (human), *Columba palumbus* (pigeon), *Acer rubrum* (red maple), *Vitis vinifera* (common grape). Linnaeus used identification to aggregate prediction, laying the foundation for a suite of repeatable horticultural procedures such as domestication, cultivation, forestation, and, ultimately, afforestation. Because achievement is at stake, normative methods prevail in order to create human agreement. This is an example of a simple application to an increasingly complicated situation and is an art of transmission that makes knowledge possible.[14] Through specialization, humans inherit a history that emphasizes economy or cultural prowess. As a result, plants remain passive, and persist only insofar as humans decide to collect, identify, number, and name them. In this exchange, there is no history of plant life. Moreover, plant life is neatly bounded, well behaved, and entirely aspatial.

The use of "type specimens" or generalizations of form was legitimized by Linnaean taxonomy or binomial nomenclature during the early nineteenth century, in the evaluation of plant specimens. A type specimen is an individual plant or animal chosen by taxonomists to serve as the basis for naming and describing a new species or variety.[15] It establishes a datum, an ideal, or a model from which to compare other expressions of the organism. While essential for elaborating the techniques of synthetic propagation, a type specimen effectively disregards varied reproductive rates, physical exceptions, unworkable behavior, and any spontaneous or wild aliveness of the plant, in favor of activating human networks. Collecting plants was an enterprise of accumulation. Behavior was nowhere in sight. In particular, Linnaean procedures evolved the type specimen as an artifact that creates the center for a dense system of correspondence and scrutiny with hundreds of emergent specialist collectors all over the world.[16]

In this case, samples were often too fragmentary to be discerned by the untrained eye, thus expertise ensued based on biologically isolated and ecologically remote interpretations of the plant that omitted exceptions. Therefore, the transmission from growing plant to static object proceeds through desiccated accounting, as most of the plant parts would have lost much of their bulk and color in the process. The diffusion at play here requires a well-oiled scientific apparatus of global connection, institutional endorsement, and specialty training in order to match a herbarium specimen with a living specimen.[17] This convoluted human association created an ordering mechanism for a new global economy that can be criticized for its imperial gaze. However, it was also a protocol that establishes the plant-as-artifact and divests plant life of agency.

Taxonomists describe a type specimen as a *holotype,* which refers to a set of selected herbaria specimens used to identify a new species. Once a holotype is designated, it is used to compare all future plant samples.[18] Omitting plants that do not conform to formal, type characteristics unilaterally imposes static measure. An explosion of expertise ensued in early botanical industries, as any reasonable knowledge of Latin enabled terms to be derived from traits, proper names to endure, and peripheral contexts to represent notable discoveries. Linnaeus named plants to secure friendships and reward colleagues for their plant-hunting achievements. For example, in *Hortus Cliffortianus* (1737), Linnaeus applied Latin surnames of people he knew, including the patron of the publication, George Clifford, director of the Dutch East India Company.[19] By immortalizing his friends, patrons, and rivals, he was able to construct a global empire of collectors who acted in the hopes of having a plant, extracted from its landscape, renamed in their honor.

Intentionally or not, taxonomy generated an exploitable resource while maturing a science. The parts of a plant were neatly arranged and ordered to accommodate the most compliant form at the time of collection: bent, twisted, and fixed from active dimensionality onto flat sheets, and into drawers. In this account, we encounter scientific knowledge rather than plant life. Such a modification of content filters the indifferent and static stuff of the world through human ideas. Thus, what allowed early taxonomists to lay claim to the title of science is not the inquiry per se but the way the discipline diagnosed power. Every plant—from the tiniest duckweed to the giant sequoia—was subject to objectification, authorized and mutilated "in the name of science."

If plants are a process that is ever-evolving, crossing, mutating, adapting, and behaving, then what is the status of the holotype? Because the holotype suggests idealized parts, the plant moves easily from the scientific realm into the physical, social, and biotic ground of the planet. The encounter with objectification remains concealed from the public because plant life typically advances unnoticed. The generic sameness of the plant-as-artifact enables planting units and reiterates across continents in afforestation schemes. Well-behaved plants, predictable plants, replace all unruly exceptions, as spontaneous development is called out as seemingly unnatural, aggressive, invasive, and even destructive. Development can no longer be charted in the same way; nor can a history of exploitation and imperialism be tolerated as the intellectual basis of environmentalism or as a given within the current strides of change across the planet.

THE THEORY OF TRANSFORMATION

"Formation and Transformation" (1807) is a short essay by the German author and poet Johann Wolfgang von Goethe (1749–1832) that summarizes his struggle to differentiate strict botanical terms from what he observed in living, spontaneous plants. The essay reads as a summary, a style characteristic of some of Goethe's shorter botanical works, those explorations that seem to be sketches of his efforts to learn about plants. Following Theophrastus, Goethe was one of the few botanists who developed an interest in how the environment affects plant life.

> When our interest in natural objects, especially the organic ones, is awakened to the extent that we desire to obtain an insight into relationships between character and function, we believe ourselves best able to acquire such knowledge through analysis of the parts . . . but these analytical efforts, if continued indefinitely, have their disadvantages. To be sure, the living thing is separated into its elements, but one cannot put these elements together again and give them life.[20]

Although Goethe is best remembered as a romantic, his publication of *Die Metamorphose der Pflanzen* (1798) or *The Metamorphosis of Plants* introduced the science of formation, now practiced as the study of plant morphology.[21] Plant morphology considers the formation of plants through development, from microcosmic activity to relationships with the environment, treating plants as a living subject outside prescriptive parts and classificatory ideologies.

Appreciating Goethean botany extends beyond his best-known works on metamorphosis because his firsthand experiments reveal a frustration with estimating plant life in parts. Goethe deftly condenses his struggle by articulating formation, the transition between forms that provoked the concept of plant morphology. The science of plant morphology was put forth by Goethe in 1790 as a means to express his own dissatisfaction with the authority of botanical science; perhaps because he was such a keen observer of nature, he could not study plant parts and put them back together again to *give them life*. From Goethe's perspective, his era of scientific research had produced a limiting dependence on measure through microscopic study and on order through taxonomy.[22] These two approaches, he believed, were quick to confirm "truth" rather than articulate and convey knowledge. In his view, organic entities achieved only temporary form, continually forming and transforming, and as a result could never be understood as being *formed*. For Goethe, the study of plants was based on firsthand experience and observation, so that even when using apparatuses or arrangements, understanding was always achieved in exchange with the living organism. This mingled exchange, or coproduction, resonates with activity rather than the measure and order of parts.

By resisting plant parts, Goethe challenged the prevalence of fragmentation embedded in identification of numbered stages. At the time, the term *metamorphosis* was used to describe discernable life stages according to an identified plan, illustrated for instance in the life cycle of an insect.[23] Instead of pursuing frozen moments, Goethe conceived of plant life as a fluid series of transitions.

Through Goethean wonder, organic entities only ever achieved temporary form, continually forming and transforming, and could thus never be understood as fully formed. Also recognizing that plants could not be forced into categories of development, he advanced a theory of the plant as a collection of slowly dividing cells, establishing the foundations of plant morphology.

I had not ceased to go forward along the path marked out by Linné upon which, however, I found a good many things holding me back if not actually leading me astray. I conscientiously attempted to apply botanical terminology to plant parts, but unfortunately was very greatly impeded in the process. For instance, when on the self-same stem I saw what was indubitably a leaf gradually turning into a stipule, when on the self-same plant I discovered first rounded and then notched, and finally almost pinnate leaves, I lost the courage to drive a stake or even draw a mere line of demarcation.[24]

This is one of many passages in Goethe's oeuvre where he recognizes that Linnaean binomials provoke a globalized classification in order to generate an exploitable inventory rather than contribute to knowledge more broadly. While he respected the requisite of industry and distribution, Goethe came to regard Linnaeus and his successors as legislators.[25] Goethe diverged from popular scholarship because he sought an alternative for those naturalists who studied plants in order to gain knowledge, not wealth and influence. Yet he sought not to replace or substitute the science of taxonomy but to only offer other paths. This is evident in a number of botanical essays whereby he articulates the distinction in intellectual bias, including "Formation and Transformation," but it was in the essay "The Experiment as Mediator between Object and Subject" (1792) that Goethe praises the long and rich history of science yet problematizes what constitutes evidence in the hands of enthusiastic scientists.

It is undeniable that in the science now under discussion, as in every human enterprise, empirical evidence carries (and should carry) the greatest weight. Neither can we deny the high and seemingly creative independent power found in the inner faculties through which the evidence is grasped, collected, ordered and developed. But how to gather and use empirical evidence, how to develop and apply our powers—this is not so generally recognized or appreciated.[26]

Goethe probes the approach to botanical studies; rather than presume the exploits and generalizations of imperialism, Goethe questions the compatibility of convergent methods, a line of questioning advanced following his Italian excursions in 1786–88.[27] An important shift in his approach to plant life materializes as a self-conscious struggle between how he learns and what he knows. Goethe could no longer adhere to a representation of the world as a resource, as he begins to write about science as an exchange between human and planetary interests. Goethe's travels between climates allowed him to appreciate the variation in plant cover as he traveled south and experienced each environmental change with a distinct shift in culture, prompting him to become more self-reflective in relation to plant life.

The change in Goethe's tone is noticeable in his prose, as he articulates nature in relation to lived, human experience. Upon occasion, this tone engages an experimental language that most scholars attribute to Goethean science that "views nature's objects in their own right and in relation to one another."[28] Goethe embraces human influence in the process of gathering data, lending agency to the relationship that exists between ourselves and other species as well as between species. Thus, Goethean science offers a remarkable precursor to the appreciation of nonhuman knowledge, historicizing the expansion of scientific outputs. An attitude of understanding and inclusion is significant, in relation to the fact that Goethe felt a correspondence with plant life. The urge to study the unfamiliar inspired methodological questions.

In the elaborate notes from his diary entries, he proposes the ideal plant, or "ur-form," as a structure or model from which an indefinite number of arrangements could be derived. Goethe speculates upon a primal plant, what he termed *Die Urpflanze,* suspecting that there was an important description that all plants share. As he records in his Italian diaries on May 17, 1787: "The primal plant will be the strangest creature in the world, which Nature herself will envy me. With this pattern (model) and the (key) code to it, one could go on endlessly inventing plants which would be logically possible even if they do not actually exist; they would not be merely artistic or poetic illusions but would have an inner truth and obligation."[29] Goethe noticed aliveness in the basic principles of growth and convergence, discovering individual metamorphoses as a sequence of effects. He observes the persistence of formation as "a continuous process of differentiation," distinguishing constant growth from static classification or the simple magnification and extraction of parts.[30] Today, his discovery refers to a developmental process called cell elongation, a process as significant to plant growth as cell division, a process that is indeed common to all plants.

By experimenting with methods beyond pure extraction of parts, Goethe observed that plants do not have a predetermined body form that simply matures in size, like a human or an animal. Thus, plant form is not genetically predetermined, which is why Goethe refers instead to formation. In this way, Goethean morphology anticipated indeterminate growth, a crucial difference between animal and plant life. Indeterminate growth has no predetermined structure or termination; it is a ceaseless development that engenders adaptation. Indeterminate growth doesn't stop, deciphered in the advance of an acorn as it becomes an oak tree, and the oak tree as it matures and spreads more acorns. *Die Urpflanze* was a model plant, a theory that describes the unfolding of repeatable modules, the tendrils that reach, and the meristem that enables plants to shed and generate new organs, changing exterior form in close collaboration with internal (genetic) structure and by means of potential environmental influence. Rather than rely on stages or parts, Goethe adjudicated the term *morphology* by applying it to a plant's ability to generate developmental arrangements. His reflections expressed how individual metamorphoses are excited internally yet endure through a process of differentiation on the outside of the plant. For Goethe, this suggestion meant that change over time could not be charted definitively as the attainment of a particular end state. Knowledge of parts hindered the study of formation.

The implications of artifact, the study of the plant in parts, are manifold because plant life cannot be put back together, physically or symbolically. This is evidenced by violent exercises like afforestation. Consequently, embedded misconceptions between outward form (the visible plant) and internal organization (the concealed plant) proliferate, further scientizing plant life. As organisms are divested of life, a subsequent decline in knowledge about the plant as a major planetary influence ensues. Botanical practices that are reliant on socialization of parts extract the plant from its context and thus its significance, which reveals that the history of systematic separation between human and plant life reduces any cooperation or shared meaning between species. In Goethe's terms, the question remains whether or not we can put these "elements together again and give them life," an appeal to synthesize spatial, biotic, temporal, and organizational dimensions.

The study of plant life endures in the twenty-first century through the tradition of generating artifacts, despite the earliest suggestions from Theophrastus and the remarkably creative science offered by Goethe. While there are other scholars and intellectuals to learn from, the reader will appreciate that the ambition of pointing to a few nonextractive achievements implies that there is a secret history of botany that awaits discovery. And it is worth the exploration as a way to offset the assumption that the only way to know plants is by virtue of their contribution to the human project. It is all but guaranteed that generating artifacts is a form of extraction, as the most visible and identifiable parts of plant life gain value and the other parts are discarded. Neither growth between organs nor the appreciation of formation provide counterweight, despite scholars being concerned with the same living world. The evidence is all around us that plant life is immensely fragmented by human systems. Can the aliveness of plants be reassessed through the fog of exploitative regimes and technological solutions? While the achievements of morphology drove a wedge between constancy and transformation in the plant sciences, this rupture only furthered the divergence between human advantage and human knowledge. As a result, plant morphology—the study of formation—belongs to a curiously minor discipline, largely trapped in a European tradition that was never promoted globally.[31] While it is possible to criticize morphology for being strictly Eurocentric or for failing to instigate a following, it is the critical power of deciding not to mobilize plant life into a measure of human achievement that sets it apart. This is also at the heart of the suggestion implicit to the politics that rest on rendering the laws of plant life calculable and predictable.

Suprapersonal Methods

Plant morphology is a subfield of botany. As a discipline of the plant sciences, it is generally disregarded outside modern botanical scholarship. Prominent morphologist Donald R. Kaplan suggests that the rejection of plant morphology as science is due to its lack of exploitative models created for the sake of economic gain.[32] If artifact suggests that breaking the plant into parts has important permanent side effects that sanction exploitation, then plant morphology is the counternarrative, for its emphasis on the study of plant life "for its own sake."[33] This sentiment resonates most radically in the scholarship of Agnes Arber (1879–1960), a plant morphologist and anatomist, historian of botany, botanical bibliographer, and plant philosopher.[34] The latitude of scholarship exemplifies how absorbed Arber was in structuring a reading of plant life; she sought to enrich and develop her studies by examining how other human scientists went about their work. Arber studied the human studying the plant.

As a significant contributor to the elaboration of plant philosophy, Arber produced English translations of various German texts while she advanced an appreciation for science as a theoretical endeavor. Arber exemplifies *plant thinking,* insisting that science was meaningless without contemplation and reflection.[35] In 1946 Arber translated Goethe's *The Metamorphosis of Plants* from the original German, introducing morphological theory in an accompanying essay that offered a glimpse into German naturalism. Her work trusted in Goethe's botany. Although reception to Arber's articles, books, and atypical publications varies greatly, it is on the basis of her creative and confident assertions that plant morphology can now be considered both a philosophy and a discipline.[36] It is noteworthy that her peers and their antiquated botanical societies rejected much of her research during her active years. Specifically, her work is criticized for its subjectivity. The questions she raises put science to the task, as Arber shows interest in thinking about *how* botany was done rather than what it was achieving, not unrelated to Goethe's search for method. As such, her philosophy criticizes the number of botanical subfields reliant on specialization, the economy of artifact.

The full significance of Arber's perspective is revealed in her translations and commentary of Goethean botany. Like Goethe, Arber enjoyed an artistic life, and she portrayed plants in drawings and illustrations as inexhaustibly as she published.[37] As her writings indicate, morphology does not conform to utility through measure or sustain capital, and resists specialized study expressed in parts.

> In these days of specialized study, the different branches of biology cannot but lead existences which are, to a great extent, isolated from one another. The aims which they pursue, and the highly technical methods by which these aims are achieved, differ so widely that one reminds oneself, with something of a shock, that all the branches are concerned with the same living world.[38]

Through her prose, Arber offers a critical voice for establishing a critique of economic botany, and elevating other, less scientized theories of plant life. The force of Arber's scholarship is found in how she contemplates the history of science as adeptly as she practices as a plant morphologist. Each of her publications opens with a chapter outlining the principles of morphology, by excavating botanic history from Theophrastus to Darwin, significantly defending the criticism that morphology endured at the turn of the century. In defining the purpose of plant morphology, she describes *form* as "determined by inner necessity," what she calls chemistry and physics manifest. Studying the plant "for its own sake" not only respects the concentrated forces of internal organization but works against the forces of capital that commodify plant life. In Arber's hands, the study of plant life necessitated a personal, reflective approach.

In *The Natural Philosophy of Plant Form* (1950), Arber's critique of specialization is forcefully rendered. In particular, Arber emphasizes three consequences of specialization: First, specialization relies on the composition of authority, which yields professional standards rather than local arrangements. Second, she explains that the influence of science is bound to a social hierarchy that excludes more than it invites. And finally, specialization expands on the presumption that plant knowledge builds historically and unilaterally on past achievements, a radical assessment she unpacks at length in *The Mind and the Eye: A Study of the Biologist's Stand-point* (1964). At each stage, Arber contends that practical or common knowledge is excluded because it cannot reinforce a contribution to the episteme. In more practical terms, this explains why land-based practices and traditional knowledge find no peer review in the system, not only because practices elude professional standards but because there is no need to produce or protect artifacts. The mingled effects of specialization inform current tendencies to lock out collaborative practices and disparage practical experience. Specialization is a prevailing account of expertise, where scientists concentrate on more and more obscure or opaque subjects that distance plant and human life, and science from the public.

Arber's morphological studies resist the artifact because she works against specialization in parts, suggesting what she calls "suprapersonal methods." For instance, *The Mind and the Eye* challenges the established techniques of specialists, surveying how scientific proof is hinged on academic endorsement.[39] This is one of Arber's philosophical texts, where she explains her frustration with how scientific experiments support the publication of words, or more specifically persist through the authority of references. Most remarkably, Arber breaks down what she calls "the biologist's problem" into five or six stages that are suggestively detailed throughout the publication. The first stage distinguishes the question, the difficulty of defining the problem. She laments that the appreciation of a valid problem is more tasking than its solution, calling for a "suprapersonal" approach instead. Problems, in Arber's mind, distance the scientist from the issue. Arber fundamentally rejects the enforcement of closed questions and objectivity, suggesting that progress might be found in the questions that are not easily answered, where no answer might emerge.

The second stage is the observation and experimental search for data, followed by the endeavor to organize and interpret, and the attempt to test its validity. Arber suggests that once these four stages are accomplished, the researcher must turn outward—toward communication—in order to add to knowledge more broadly and reach beyond specialization. Arber suggests that research methods can reach outside the dominant paradigm, as she proposes a sixth stage in the sequence.

> In this the biologist stands back from the individual jobs to which he has set his hand, in order to see them in the context of thought in general; to criticize their presuppositions and the mode of thinking which they employ; and to discover how the intellectual and sensory elements, which they include, are interconnected.[40]

The plea to gain perspective outside the scientific community is immanently public and social, as it is only achieved once the question is tested, in the fifth stage, within the scientific community.

From this point of view, it is interesting to consider the critique of Arber's work in light of current scientific mistrust due to corporate entanglements as well as to the loss of common practices that are deemed unproven and unscientific. What if scientists were beholden to public outreach as systematically as they are held to peer review? I ask this question in light of how little the public knows about plants, and seemingly, how isolated botany can feel from the outside. Arber's challenge to her discipline resonates not only within science studies but in more direct and practical ways in the collective plea to slow the sciences.

Consider the articulation of Arber's first stage, which concerns the formulation of the "question." If the inquiry into environmental decline counters with afforestation, then questions are designed to advance simplified, singular solutions. Such solutions yield commodified analyses, including the purchase of specialized equipment, spaces for controlled experiments, and mechanically advanced regulators to generate nothing but outcomes with clear transferability. As scientists are encouraged to ask "well-posed" questions, remedies are designed to point to the "right solutions." Beyond biotechnology, laboratory selection, and sponsored aid programs, what is at stake in this culminating process is how the manipulation of biological systems circumscribes aliveness. It also has very little to do with what Arber calls "far-reaching problems" or "suprapersonal questions." Thus framed, afforestation is a product of abridged problem statements.

Static evaluation yields vital insights to the science of knowing how plants can serve humankind, but without scholarship for the sake of plant life, the living plant cannot translate back to environment. This has tremendous repercussions as it relates to the rise of specialization in twenty-first-century biology, which now monetizes tree planting through ecological collapse, deforestation, and species extinction, for instance when amassing biofuel acreage or gathering carbon credits. The articulation of suprapersonal methods helps conceptualize the act of planting as the ultimate engagement between plant and human life.

Artifact empowers specialization and neutralizes practical experience. Of particular interest to afforestation is an account of plant life that does not obey, one that resists parts and altogether cannot be categorized and exploited as effectively. Arber's suggestive methods expose the attachment of static form and scientific certainty. The contributions of Theophrastus's development between parts found no place in the science of taxonomy or the conquests of economic botany, nor did his attention to context and the wild side of ecology. Goethe's morphology in prose was not scientific enough to be applicable to the material, rational world. The plant's most salient powers of growth and collaboration are disregarded, as parts find priority over the processes that unite them. By marginalizing the living plant, and removing all associations with physical context, any ability to produce future generations becomes a specialization. The account of misconduct is exemplified in this section by *Faidherbia,* a plant that resists exploitation and complicates single resource harvests, further confusing economic decisions. The proposition is that the planted world would look very different if a less-specialized aggregate of humans was allowed entry into the discoveries between intellectual and sensory experience, whereby knowledge is refreshed by the formation, disobedience, and behavior of plant life.

Life in Evolutionary Terms

In order to overcome the plant as artifact, imagine four billion years of plant evolution by picturing an ancient earth occupied entirely by microbiological organisms. Each mat-like community blankets its oceanic or rocky habitat, emerging as tiny forests of bacteria. No roots, no stems, and no leaves were invented yet. Most of these microorganisms are heterotrophs—they require organic matter for nourishment—so they float on the surface of the ocean and feed. They do this so successfully that they create a food crisis, consuming all available organic material. At the same time, some microorganisms appear to be autotrophs, producers capable of metabolizing inorganic substances into organic ones and essentially feeding themselves.[41] Bacteria are prokaryotes, or nonnucleated cells. Organisms with nucleated cells are called eukaryotic. This "split" in the microorganism world is considered to be more exceptional biologically than the perceived split between other kingdoms, casting "life" beyond simple definition.

> Life, although material, is inextricable from the behavior of the living. Defying definition—a word that means, "to fix or mark the limits of"—living cells move and expand incessantly. They overgrow their boundaries; one becomes two becomes many. Although exchanging a great variety of materials and communicating a huge quantity of information, all living beings ultimately share a common past.[42]

For instance, plants, animals, and humans share more common ancestry than prokaryotes and eukaryotes because our cells are membrane bound and contain nuclei. Although they are in the minority, autotrophs prosper by metabolizing water, sunlight, and carbon dioxide, releasing small amounts of oxygen in return. Autotrophs flourish because they are independent. Although microorganisms such as these are very small in size, their functioning engages all the same processes as more familiar organisms, including humans.[43] We could not exist without this ecological recycling, linking creatures of all sizes both metabolically and ancestrally.

Autotrophs survived the exploits of early wave action, as ancient seas washed them ashore onto rocky, unsteady shoreline surfaces. Heterotrophs required proximity to water and organic material, a foot in both worlds, so to speak. Over millions of years, autotrophs gathered and reproduced by clinging to the coarse, barren surfaces of the planet and feeding themselves, averting the food crisis created by the heterotrophs. Fast-forward another billion years and there is a lot more oxygen in the atmosphere because these tiny autotrophs consistently released small amounts of oxygen over a very long period of time. This is an early example of how life both changes its environment and then adapts to the changes, which in turn alters the environment enough to demand further adaptation.[44] An evolutionary process is a biological necessity, and in this way, organisms play a part in their own evolution.

Microorganisms conditioned their air, transforming the environment by producing oxygen. This transformation facilitated the rise of oxygen-breathing species and still does. An appreciation of tiny, nucleated cells is fundamental to a full description of life, offering non-hierarchical and inclusive insight into human kind. Studying microorganisms and nonhuman agency might just make us better at being human. When Lynn Margulis (1938–2011) reiterates that "all living beings share a common past," she is not describing separate vectors of evolution, an imaginary shared past of human ancestry separate from the trajectory of other lineages such as those of plants and animals. There are no colored parallel lines that separate and detach. Margulis is describing only one vector or drift of life through the lens of conscious bacteria. In describing our common ancestors as "two that becomes many," Margulis emphasizes that bacteria and other microorganisms actively shape the earth. It would be a mistake to historicize her statements because she is expressing active forces. The suggestion is especially compelling because, in her words, "species specific arrogance" and the "tenacious illusion of special dispensation" limit humankind from accepting that all species, including bacteria and microorganisms, consciously create their own conditions.[45] Margulis's research is motivated by an aspiration in evolutionary biology to comprehend how biomass emerged, in order to argue for our place in earth's history, which might be surprisingly minor.

Time and space are brought into discourse, as genetic drift drives the recognition that no living body is ever static and that humans and plants share the same bacterial lineage. Margulis positions the primary driver of evolution by "symbiogenesis," a concept that significantly advances the conventions of evolutionary theory. Symbiogenesis is an evolutionary theory of early organisms, first described by Konstantin Mereschkowski (1855–1921) and substantiated by Margulis. Symbiosis simply means living in close contact, but in Margulis's terms some organisms do not just live together but fuse communities and create a third kind of organism.

As members of two species respond over time to each other's presence, exploitative relationships may eventually become convivial to the point where neither organism exists without the other.[46]

Thus, bacteria evolved into protists, and subsequently evolved into multicellular beings through shared determination, by splitting up and getting back together in a remarkably productive relationship. According to Margulis, symbiotic mergers are responsible for transmitting and inheriting variation along the way. The concept of individual mutations is less relevant than a consideration of exceptional assemblages that she calls "consortia," communities of organisms that join and collaborate through life. In each study, Margulis returns to the recurring example of lichen to demonstrate her articulation of symbiogenesis. Lichen is an organism that is neither plant nor animal. Lichen is composed of one-part fungi and one-part photosynthetic cyanobacteria, forming a partnership that intertwines to become a unit. The lichen individual is "life" distinct from its component parts: "The result: an altogether new life form that takes advantage of the alga's ability to make its own food and the fungus's ability to store water and fend off the elements."[47] Thus, lichens can cycle between extremes in mixed light and moisture conditions, where either its algal part or its fungal part manages growth. Lichen is a state of being. Lichen existence is found in its quick microsurvival that rests on the apparent individual, so that each organism can take part in growth, truly an exemplar of living together.

Early autotrophic plants clung to shorelines and grew close together, forming patches along a gradient of moisture. Hugging the rocky shores enabled slow landward evolution, trending away from their aquatic dependence. Rather than competing with one another, they formed symbiotic relationships within patches and amid the extremes of a shifting atmosphere. Tiny rootless plants colonized the rocky surface, creeping along a soilless horizon. Plants evolved with other plants, with rock, with water, with the chemicals in the climate, and without humans or other creatures. At this time, estimated at eighty million years ago, tolerance to moisture and drought helped advance the slow colonization of plants across the surface, as symbiotic relationships were not limited to these interspecies exchanges. Each modification and advance implies that plants invent their habitat as they move or grow; in this case, generating a physical environment with less carbon and more oxygen became a useful condition.[48] Early autotrophs invaded land by forging relationships that fostered spread and growth. Survival on land was not isolated; it relied on a sophisticated aptitude toward "living together," ultimately creating conditions for all other life forms, including humans. Because this single lineage of plants learned to skillfully work with others, they flourished where others failed, spreading across the planet.

Plant life spread by either reproducing in water (gametophytes) or using wind by evolving drought-resistant spores (sporophytes), borrowing small breezes or riding on large storms to gain ground, in much the same way as they had increased in the water with surge and swell. At the time, air was just a new medium for the eager plantlet. Over time, growth evolved vertically, as the recognition of this new airborne energy source was revealed, as plants recognized the value of elevated stems and allowed their spores to meet the breeze and travel further.[49] Terrestrial plant life collaborates with the wind. Even in the most extreme environments, inputs create and increase reproduction, a behavioral collaboration that grew to include humans and other animals. Movement and spread evolved in tandem, proliferating territorial dominance.

As new inputs create and increase reproduction, it is possible to imagine that plant life is evolving further, with more recent incorporations of augmented anthropogenic atmospheric encounters, fewer pollinators, and the hotness common to impervious surfaces. It is also possible to imagine now that plant life works with the elements, and that the environment— naturally evolved or anthropogenically crafted—creates a productive encounter for plant life. The question is only which species adapt and which ones fail. What is also humbling is that encounters progress with or without human life.

Productive symbiosis summarizes life in evolutionary terms. Within the extremes of devastating desiccation, tremendous solar radiation, and nutrient scarcity, plants dramatically restructured the earth's surface.[50] They did so by working within partnerships that tended toward collaboration. Each relationship is inconspicuous over timescales that are tacitly out of sync with our species. The serious temperature extremes in which plants took early advance completely outstrip current temperature anomalies that are causing humans to develop theories of plant conservation, diversity, and bioengineering. By conserving extant plants, we are disregarding the adaptations, behaviors, and migrations that cannot be conserved. Acknowledging that plant life advances by letting other forces and organisms into their radical consortium puts pressure on "end of the world" thinking and instead reduces the human to just one of those forces. If human temporal and spatial partnerships continue to insist on treating plant life as an insubordinate diversion or a productive tool, then we will remain fixed in our ways, compelling the artifact, and advancing out of sync with plant life.

2

Great Green Wall

THE GREAT GREEN WALL for the Sahara and the Sahel Initiative (GGW or GGWSSI), or La Grande Muraille Verte, is a comprehensive supranational, cross-continental tree-planting plan to consolidate the ecology and economy of eleven countries from West to East Africa. The initiative is packaged as an environmental solution to an environmental problem. Afforestation advances by the same botanical authority that—intentionally or not—generated an exploitable resource and reduced the aliveness of plants. The artifact is relevant to the African case because it emerges in the past so convincingly that it slips into the present. Artifact is contemporaneous with the currency of tree-planting units.

Using mapping as a device and desertification as a label, the Great Green Wall mobilizes the artifact of botany in an effort to associate vastly divergent landscapes: "Each country concerned with the wall should provide a strip of 15km wide in compliance with the general indicative course. The strip should necessarily be located within the Sahel zone with average rainfall between 100–400 mm. The green wall would thus be a large green avenue, more or less linear but continuous and as far as possible."[1] The ways in which the initiative advances the technique of planting individual trees across the continent makes explicit that dryland uncertainties— erosion and drought—are the only common motivations for the project. Extant plants are overlooked, as is their behavior, and thus their potential to engender reciprocity within the cultural and biotic landscape.

What is striking about the GGW is how the distinctive regions between the Sudano-Sahelian savannas and the Sahelian shrublands is covered up by a singular response. Problems of unequal landholdings, fuelwood shortages, and nomadic pasturelands tend to lie outside afforestation policy. The careful sidestepping of difference simplifies the response, advancing a tree-planting campaign that boasts continental scales, cross-border ambitions, and advances in technology provided by "high definition vegetation mapping."[2] Nongovernmental agencies play the role of developer, manipulating the rules of the game by establishing a remote relationship to the land under consideration. The "field" is mapped and remains "out there," a landscape rendered powerless to the incoming plantation. Trees are planted but "owned by nobody."[3] In a strategic bending of the rules, plants are drafted into service as environmental tools to cover up social injustices as tree planting emerges as a trusted environmental "good." Through the lens of artifact, political judgments pair with international procedures, in spite of the unjust and simplistic generalization of drylands.

The project for the Sahel–Sahara region was officially proposed in 2005 by the Republic of Nigeria's president, Olusegun Obasanjo, and promoted as the only solution to encroaching desertification. The project anticipated further implementation strategies to satisfy the request made by the United Nations Convention to Combat Desertification (UNCCD) in Rio de Janeiro at the 1992 United Nations Conference on Environment and Development (UNCED, or commonly the Earth Summit). A year after President Obasanjo's announcement, the project gained further momentum during the Conference of Parties to the Climate Change Convention (COP 12) in Nairobi, Kenya, in 2006. There were no conclusions or commitments agreed upon during COP 12, but there was one noteworthy consequence: it was the first COP held in Africa. Introducing the reality of dryness and the experience of aeolian transport to specialists traveling from humid countries secured support and capital among an international community of governmental and nongovernmental agencies. In order to sustain a single resolution across the inconceivable variation of Sahelian sociocultural and biotic landscapes, the promoters of the Great Green Wall endorse a replicable solution: tree planting. Tree planting is justified by the problems of environmental decline and helps maintain an image of verdurous, profitable, and safe settlement. Deciphering who is soliciting the question, and why afforestation is the answer, is the challenge presented by the African case.

The list of agencies involved in the Great Green Wall project is a complex and multifaceted array of international, pan-national, and national organizations, all of whom are amalgamated into one five-letter acronym: NEPAD (New Partnership for Africa's Development). The partnership provides regular public updates that often reaffirm its ambitions.

> The Great Green Wall project involves planting a living wall of trees and bushes more than 7,000 kilometers long and 15 kilometers wide, from Dakar, Senegal in the west to Djibouti in the east, to protect the semi-arid Sahel region from desertification.[4]

Although the GGW project is still in its formative years of determining sites, preparing planting lists, and initiating pilot projects, NEPAD offers a continually worked and reworked document that mimics escalating global concern and shifting regional markets. Claiming to lead the project in partnership, the Community of the Sahel-Saharan States (CEN-SAD) and the African

Union (AU) also created the Pan-African Agency of the Great Green Wall (PAGGW) in 2011. For policy purposes, each country is responsible for adopting and implementing plantings on their respective soils. Funding is secured in sizeable amounts from various sources: for example, hundreds of millions of dollars from the World Bank, Food and Agriculture Organization of the United Nations (FAO), and the United Nations Economic Commission for Africa (UNECA).

The number of countries involved in the project obscures the possibility of gauging total capital investment and its associated distribution. Government organizations receive international funding for planting, but the complicated network of agencies makes it difficult to measure effectiveness, if there is any at all. With eleven national governments and countless nongovernmental organizations subscribed, the initiative is an experiment in geopolitical collaboration, not to mention the implicit cultural, linguistic, and social discrepancies aspiring to cooperation. The clarity of the championed solution, "plant trees," reduces action to a symbolic image. Artifact resurfaces as each stage of increased specialization dismembers the plant in order to increase the legibility of an initiative without a project. Despite the layers of funding, the marketing campaigns, and the promises of redacted species lists, there is little to no evidence of project making to date—the Great Green Wall is a well-funded committee.

An exploration of the GGW highlights a twofold situation: First, project-making efforts of the GGW engage the lineage of colonial forestry across the Sahel because the project does not propose anything novel; it only reiterates the breach presented by foreign tendencies. Programs and policies are administered through the use of desertification as a strategic problem that positions afforestation as a solution. The approach is a static, long-standing interventionist strategy to create a debate by those who hold power and typically live outside the affected areas. The second situation also evolves environmental decline narratives, this time by the international aid community, mobilizing piecemeal improvements to prompt steady, organized action. Humanitarian elite enable the state to regain control of marginal lands. The attention to "tree planting" is so deeply embedded in the inadequate relationship between plant and human life that empirical proof, statistical models, and persuasive specialists are tendered as corrections to traditional knowledge. The artifact reiterates as climate agendas replace desertification agencies, extracting practical and land-based knowledge to disseminate reliable, replicable proof that afforestation is both the best policy and the most effective practice in drylands.

INVENTING *DÉSERTIFICATION*

The term *désertification* is not an ecological term. It does not describe a geologic, climatic, or biotic process. It was invented in 1949, proposed by a botanist reporting for the French Forestry along the south coast of West Africa. At the time, L'Afrique Occidentale Française (AOF, French West Africa) was the colonial and administrative territory of French oversight whose makeup changed over time from 1895 to 1958. AOF included what is now the Republics of Guinea, Côte d'Ivoire, Mauritania, Senegal, Mali, Niger, and Burkina Faso. André Aubréville was hired to account for a drop in yield in this context, a region of subtropical Africa categorized by 700 to 1,500 millimeters of annual rainfall, as opposed to the approximate range of 100 to 600 millimeters across the Sahel. Seasonal mean rainfall is largely driven by large-scale atmospheric circulation, which accounts for the expansive range. For instance, current models suggest that the Sahelian range varies from north (100 to 200 mm) to south (500 to 600 mm); either way, subtropical rainfall far outpaces those inputs across the Sahel.

Aubréville was stationed in the humid, verdurous Côte d'Ivoire, writing and reflecting on *la fôret colonial*—dynamic old-growth forests that were being harvested for timber. When he published on *désertification* for the first time in *Climats, forêts et désertification de l'Afrique tropicale* (1949), Aubréville was specifically referring to the transformation of old-growth forested land to unfertile land, insinuating that the effects of industrial clear-cutting were multiplying in humid regions: "These are real deserts that are being born today, under our eyes, in countries that receive 700 to 1,500 mm of rainfall."[5] He describes how under the pressure of human influence, tropical rainforest can be transformed into savanna, and savanna into desert.

> You do not have to be a visionary to perceive the unmistakable image of Africa's future. . . .
> At the biological scale, Africa is tending toward savannah—a naked savannah.[6]

At the heart of *Climats, forêts et désertification* lies a description of how colonial forestry policy was accelerating soil scarcity; harvesting too many trees ruins the topsoil, a condition to which the forests of subtropical Africa react by desiccating severely. Identifying a shift in paradigm was certainly one achievement, but Aubréville went one step further, naming the condition and thereby giving it identity and consequence: "desertification," he claimed, results when desiccation, accelerated by human exploitation, manifests itself during a period of low rainfall.[7] By adding action to the noun *désert,* Aubréville proposes a verb that acknowledges the movement and transitions inherent in ecology and manifest through disturbance. Aubréville went on to add that human action was not only damaging ecological quality but also leaving land vulnerable to erosion, a condition from which it could not spring back.

The idea of desertification was not new when Aubréville introduced the term. In France, the naturalist and colonial administrator Henry Hubert coined the term *dessèchement* in 1920.[8] The following year, the historian E. William Bovill recorded diminishing rainfall in Nigeria and put forward the observation as evidence of what he called "the encroachment of the Sahara on the Sudan." His claim was never scientifically substantiated and, additionally, lacked a catchy name.[9] The most energetic protagonist of the "encroaching Sahara" scenario was the British forester Edward Percy Stebbing, who was alarmingly eloquent when describing the escalated threat to colonial land holdings in West Africa, using the term *desiccation* to describe the process of dehydration or dryness resulting from the removal of water.[10] Stebbing also referenced the American soil conservation movement and the recently established U.S. Soil Conservation Service (1933), a federal agency, both of which voiced the consequences of imprudent land practices, such as large-scale deforestation and slash-and-burn techniques inherent to industrial activities. Stebbing was a great influence on Aubréville, a fellow botanist who mistrusted the motivations of colonial forestry and the escalated agricultural cultivation it stimulated.

By proposing the term *désertification,* Aubréville helped construct a narrative that would resonate further than he could have imagined. At the time, the practices of French-held West Africa were exemplified by forest policy, positioned as a means to secure profit through resource extraction. Once in possession of large tracts of land, French colonial oversight created two systems for administering the land under their control: classifying land type through mapping, and adopting specific language for labeling unproductive conditions. The active, mobile, and therefore threatening movement of desertification engendered the perception of dryland decline and the rise of environmental authority. Diana K. Davis refers to this perception as "desiccation theory," a concept that widely circulated the interpretation of deserts and drylands as ruined landscapes.[11] The term *desertification* exemplifies both of these responses, as units could be mapped and labeled at once.

In *Climats, forêts et désertification de l'Afrique tropicale,* Aubréville demonstrates order through the obligatory comparative charts, or indexes, but—more critically—he drafted elevation drawings to represent and characterize the interdependence of plants. These drawings reveal associations above- and belowground, exploring vital details of plant life and acknowledging, for example, the herbaceous (nonwoody) layer and the ligneous (woody) layer, which are distinguishable as interdependent in his drawings. He recorded the behaviors of plant life, marking wind direction and the presence of birds, insinuating seed dispersal patterns and reproductive spread. As a result of fieldwork achieved close to the soil, direct experiences with living plants and an intimate engagement with the micro-conditions of the forest, the young botanist was able to submit a detailed perspective of plant life, offering more than a marketable commodity.[12] Trained as a botanist, Aubréville recognized that the forest fluctuated as a result of a shift in species abundance instigated by disturbance and exploitation, namely colonial forestry.

The message of desertification was that nature was not a divine and endless gift, a significant offense to French forestry officials who initially refuted Aubréville's claim as an attack on their science. Instead of being rejected outright, however, the term was adopted to describe any dryland condition, and to argue for its yielding to human intervention. When applied to the treeless desert, the threat was resolved by planting trees. The term was finally adopted into common use when this particular correlation between ruin and salvation could be made or was accentuated by the forestry commission. While Aubréville is known for having expressed the influence of climate on biotic abundance, it was the colonial foresters who exploited one component of his thesis: that tree cover could mitigate the relationship between the soil and the climate.

By identifying *le probléme de désertification* in 1949, Aubréville raised concern within colonial forestry traditions. The notion that the soil was not resilient to exploitation, or that overcutting and burning forest trees accelerated the condition, led to a series of new counteractions. Typical responses projected improvement strategies and logistics that championed the planting of more trees, to counteract the effects of desertification. Significantly, these strategies were not suggested to counteract deforestation. Tree planting would eventually be introduced as the victorious rival to desertification, as illustrated by Africa's GGW, which exemplifies the universal repercussions of casting plants as a solution to exploitation. This is where the power of *artifact* emerges: in the radical reclamation of scientific fact, predicated on establishing a relationship to "real world" problems.

The appropriation of desertification is simply one example of the purposeful application of ecological principles to highlight apparent environmental disorders. So-called technical terminology relies on intentional vagueness and results in overuse and maladaptive outcomes exemplified by afforestation. And terms propagate more terms: desertification can be countered by another contemporary label, "greening." An invented ecological paradigm occurs when disciplinary expertise yields to the need for uncomplicated meaning. Desertification is especially problematic, as few definitions proposed for the term are explicit about its process, such as where or when it is likely to develop, how it spreads, and whether it is a permanent or cyclical condition. The concept itself is generally misunderstood and has been commoditized and popularized beyond recognition of Aubréville's meaning. According to Helmut Geist in *The Causes and Progression of Desertification,* there are more than one hundred existing definitions of the term.[13] The UNCCD controls the current definition, asserting the condition is "land degradation in arid, semi-arid, and dry sub-humid areas resulting from various factors, including climatic variations and human activities."[14] In the current situation, where endless acronym-enhanced agencies fund planting projects, their ambitions are shielded behind so many layers of antiquated principles that it is difficult to imagine, as Arber puts it, that "they are all concerned with the same living world."

THE NATIONALIZATION OF TREES

The artifacts of colonial forest code create repercussions on the ground because tree-unit calculations obliterate human–plant stewardship. In the Sahelian milieu, afforestation fails to recognize that most trees in the Sahel are borrowed and lent based on an exchange between seasonal needs. Trees are not individually owned; they are commonly valued and shared.

> The trees in the landscape are there because they have a value to the local human population—whether for shade, wood, food, medicines, fibers or spiritual reasons. If this pattern of land use has long been practiced, why is there a need for development foresters or other experts to promote agroforestry? If the rural peasant already values trees, why does she or he need to be convinced that trees are good?[15]

The French colonial government discouraged tree felling and pruning in order to maintain valuable tree products, establishing laws that made all trees state property.[16] Separating acreage was achieved under an imperial gaze, a violence to the landscape and those that depend upon it. Dividing forests and agriculture facilitated the exploitation of forests, accounted for by a vast scholarly literature on the effects of colonial rule. Of relevance here is that forests continue to be exploited; although nongovernmental authorities contend that African deforestation is abating, it is important to point out that they do so in comparison to other "global" means.[17] Climate models are also correlating African equatorial deforestation with lower rainfall patterns across the entire continent.[18] The intensity of the change is found in the margins of the sub-Sahel, where millions of acres are lost in Eastern Africa and the Congo basin, the world's second-largest tropical forest after the Amazon. The same global influence that designed the image of the encroaching desert by popularizing desertification suppresses the anxiety that deforestation causes major hydrologic instability. Despite the typologies of planting inherent to establishing trees in drylands, no mechanisms have been set for compliance with forestry standards when it comes to taking down trees.

When trees are legislated by the government in the twenty-first century, policy reiterates colonial abuses and class-differentiated access. This raises significant concerns over how tree-planting projects are integrated within cultural contexts. Blurry and seasonal ownership lines are hard to refine in high-definition mapping. So too is any meaningful resolution of woody plant life that can take years or even decades to bear fruit, reduce erosion, or produce resin, for instance. The nationalization of tree units is a significant contradiction to how family labor cares for trees in drylands because trees grow shared resources and food supply.[19] It is not a matter of explaining value to rural peasants; it is a matter of reestablishing value at the scale

of the individual tree. The trajectory of Africa's "development" overlooks the underlying relations of colonial policies that continue to produce unequal tenure regimes.

Outside development administrations, "participatory" and "community-based" practices
still do little to dismantle the colonial policy, as forest programs continue to benefit the commercial elite.[20] For instance, lands in Burkina Faso with significant tree cover are fenced off
from community access. Once identified, these *productive* areas are immediately transferred to
the state for "forest and forest products," consequently penalizing unauthorized entry.[21] The
policy creates a public and private domain, serving powerful economic interests alone. Land—
symbolized by the presence or absence of trees—is called out, necessitating state intervention
and specialized protection. Appeals to conserve *productive* land repress the essential human
infrastructure that cultivated and cared for the trees in the first place, leaving indigenous claims
to trees out of the code. The trees no longer belong to those who plant, cultivate, and tend to
them. In turn, restricting access to treed landscapes by rule of law amplifies the value of annual,
edible plants because fields (not trees) can be safely owned or managed by a village without the
fear or threat of the land being appropriated. The ignorance toward indigenous claims is exemplified by the story of the Majjia Valley afforestation project, an endeavor that is explored later
in this section.

A simple change in policy would take years to proliferate across the landscape, not only
because of the immeasurable impact of legal frameworks but because so much tree planting
is now predicated on sedentary lifestyles that promote individual ownership. This particular
settlement tradition will be explained further in the second section of the book, within the
American case. For now, consider how access to land and tenure rights are complicated over
time, as households endure colonial patterns and emerge into another derivative form of political authority. Global development authority is hard to delaminate from past abuses. Moreover,
precolonial traditions that respect the mingling of matter and give-and-take relationships have
vanished, and so has the memory of subsistence farming because of *global* demand. The specificity of colonial forestry erased lively relationships, buoyed by the economy of timber and
trade in high-yield cash crops, creating a separation in land use between forestry and agriculture that endures to this day.

The achievements of nationally owned tree units erase any local participation and reveal
household reluctance to participate. Adversity toward tree planting generally endures throughout a family because policy spans generations. This reluctance often stalls afforestation campaigns. Why plant a tree—or cultivate and tend to it—if you cannot benefit directly? Thus, the
inheritance of forest policy moves from the tree unit to the community village. Subsequent to
the reluctance to participate, the decline of tree cover is neatly attributed to local neglect rather

than the waves of poor policy that motivated villagers to stop tending their trees, for fear of the land being stolen by the state.

The lack of "local" knowledge or the conviction that "education" is lacking is a target for development aid, designed by exclusionary policy. Blaming the lack of specialized knowledge on the small-hold famer is extended by the private, nongovernmental, and governmental agencies that proliferated following the Sahelian drought of 1973.[22] In turn, villages are blamed for the lack of material relations between plant and human life. This opens the door for global expertise, which appears on the ground through specialization from external funding agencies. Curiously, the simplified problem statement leads to an equally generalized response: tree planting. The instruments of authority continue to pacify plant and human relations on the ground, popularizing greening and planting without consensus on the impact of colonial traditions: for instance, independence from France is actually commemorated in the Republic of Niger by planting trees, as the influence of American Arbor Day sneaks its way into global politics. More recently, the day was renamed Fête Nationale de l'Arbre as an expression of dedication toward fighting desertification and greening the desert. There was no mention of the trees that were once stripped from the land, just as the disruption to common resource sharing and land-based traditions remains silent. The presumption that *productive* land can be serviced by global specialists is a web of belief that discounts the relevance of traditional knowledge and reiterates a colonial assumption that science cornered the market on truth.

Global excitement and fervor for planting trees even replaces planting trees *in* Africa with planting trees *for* Africa, entirely disregarding context from the rhetoric. As long as trees are being planted, location is subordinate to statistics. The GGW is set apart from the other cases because jurisdiction is uniquely global. Funding sources range from the World Bank to Google, raising questions of value along with questions of scale. Paying attention to the common imprint of land tenure and single-export crops is essential to an assessment of how to recuperate landscape practices. For any future afforestation project to be a "success," land tenure, ownership lines, numbers planted, and allocation rights will not suffice because they are designed to satisfy plantation cropland alone. Can a tree commons provide a framework to help rebuild human and plant relationships rather than satisfy the counts and tallies of global markets? How might policy engender collaboration?

New Forms of Extraction

It takes an impressive imagination to account for how organized tree-planting projects are introduced into contexts so varied that they span millions of square miles. It is equally imaginative to coordinate action across a continent on issues as diverse as land degradation, food production, biodiversity, resource management, community livelihood, and carbon emissions. Descriptions such as a *mosaic* of natural resource agendas creatively distract from other, more overt uses of language: accelerate, tackle, invest, manage, mitigate, improve, build. The metaphor of the *wall* likewise offers a clever way to proffer similarity and generate significance across distinct regions beset by diverse concerns, despite the effort of project statements to diversify the ambition by turning attention to community livelihood as a business opportunity.

> The Great Green Wall initiative is hosting an international forum to build new partnerships and accelerate progress in tackling one of this century's defining development challenges—land degradation, desertification and drought. The forum will also take stock of the achievements and future challenges of the Great Green Wall for the Sahara and the Sahel Initiative. The two-day event, organized by the Global Mechanism of the UNCCD (GM) and FAO under the auspices of the African Union Commission (AUC), is taking place on 16–17 December at FAO Headquarters in Rome.... Bold coordinated action and more investments in sustainable land management are needed to boost food production, help people adapt to climate change and mitigate its effects, support biodiversity, enhance businesses based on land resources and contribute to a green economy. A mosaic of natural resource management programmes underway in some of the countries, demonstrate the potential of sustainable land management to boost food security, improve community livelihoods and build the resilience of the land and the people to the changing climate.[23]

The Sahel requires an overt model of appropriation, one that can be likened to extraction. The extraction procedure relies on generating "global" attention in order to activate "aid." In order to do so, donor agencies withdraw evidence from small-hold farmers, foregrounding tree planting as a form of support. Afforestation advances through the conviction that global support is crucial to Africa. As an indication, the GGW has received $108 million from the Global Environment Facility and another $1.8 billion from the World Bank alone.[24] Such staggering figures give rise to the question of who controls the funding and how it is distributed among the various African entities that claim to be "leading" the effort, including the Permanent Interstate Committee for Drought Control in the Sahel (CILSS), NEPAD, CEN-SAD, and the African Union.[25] Since there is no spatial evidence of the project, spending is entirely obscured. Where is the money going? Evidence only accrues through the few "success" stories that are common to all accounts of the GGW. If evidence accumulates based on a few unique exemplars, then how do those paradigms stand up to replication, typology, or other standardizing mechanisms in order to satisfy global agendas? The appropriation of "success" stories can be likened to a form of extraction, as localized efforts are appropriated to conceal the failures of international development.

A closer examination of the most overcited projects exposes the disparity. For instance, despite the eleven countries involved in the GGW, the techniques of decentralization are repeatedly mentioned in Burkina Faso, Niger, and Senegal, falsely extracting evidence of the GGW or proof that the money is actually being spent on tree planting. In Senegal, claims by the same press release to take stock of the achievements of dryland planting boast that twenty-seven thousand hectares of degraded land were restored by planting eleven million trees. These statistics emerge from one of the first failed pilot programs of GGW. A pilot program is a spatial typology that repeats across afforestation projects using analogous procedures: import seed stock, establish a nursery, employ local inhabitants, and establish "tree blocks." Across the sub-Sahelian range, each parcel is fenced off from villagers, which explains the household reluctance to participate. According to a herder in the area, Amadou Issa Sow, "people said that this project was going to decrease the amount of pastureland too much," a sentiment that resonates on the ground as people cut through fences to cross the area that had been set aside.[26] Here, statistics are subordinate to practical outcomes and actual livelihoods.

Techniques are also extracted to provide evidence for future funding. For instance, in Burkina Faso, the traditional use of *zaï,* a technique to dig or build shallow pits that collect and concentrate rainwater, amplifies yield, subsequently informing policy and analysis.[27] *Zaï* are traditionally used across the sub-Sahel, and are colloquially called pits. When rocks are collected and stacked, this technique is also known as a stone *bund*—a *tassa* in Niger or *towalen* in Mali. Each pit is dug approximately one meter apart, but each pit tends to vary in size, scale, and spacing. The differences in size are likely based on soil conditions and the competence of the crop being planted. These are metrics that cannot be replicated or standardized. Despite the variability, *zaï* pits are measured and monitored across drylands because of increased yield, associated with climate-smart agriculture (CSA). The extraction reveals *zaï* pits as a quantifiable technology, driven by the collection of evidence.

The potential scalability of *zaï* pits endures by extracting suprapersonal knowledge. When a small-hold farmer, Yacouba Sawadogo, began to add manure to his *zaï,* he noticed that trees began to sprout thanks to the seeds embedded in the manure. By passing on his experience to others, he instigated a wide appropriation of a basic, primeval tradition: careful fertilization. Fertilization amends and augments local soil conditions. As Sawadogo's exertions are appropriated by numerous eager correspondents, research papers and international policy shift to emphasize the role of "natural regeneration" for climate readiness.[28] One of the most shockingly practical questions to ask is how or why knowledge of local soil improvements was lost.

Another question is how its recovery was extracted to fill a gap in the *mosaic* of techniques that promise to alleviate chronic food insecurity in West Africa.

Natural regeneration is not a technique that requires trees to be planted. It does not activate the progression of experts in species selection, plant importation, or nursery establishment. Natural regeneration acknowledges the time it takes for plants to sprout and grow with minimal human effort and attention. More importantly, *zaï* pits shape a careful association between the farmer and the soil. Nurturing *zaï* pits ensures that a relationship is established between those digging the pit, the manure being mixed in, each soil microbe, and the saturated dormant seeds. The story of Yacouba Sawadogo is adapted to represent the aforementioned goals of Africa's Great Green Wall in "boosting crop and livestock yields as well as the production of medicine and firewood," but the most critical metric is omitted: the project to enlarge the practice of digging *zaï* pits started in 1985. It is only over three decades of careful cultivation and human attention to the dynamics of the landscape that trees are emerging. This has nothing to do with technical innovation, conservation policy, or natural resources management and everything to do with patient collaboration between plant and human life.

The visibility of *zaï* pits and bunds materializes the landscape of "aid," as the dryness of the soil is disguised by the scarcity of trees, or irregular stone piles obscure the regular intervals of development technology. The *zaï*-bund pilot project is insufficient if it does not integrate action. The supporting material in this case, offered by Resilience through Enhanced Adaptation, Action-learning, and Partnership Activity (REAAP), reinforces visible action: "This conservation work will be re-enforced by planting tree seedlings from community nursery sites to increase community resilience to climate related shock." Much of this message is found in the metadata associated with a photograph but is reiterated on the website of REAAP, a project led by Catholic Relief Services (CRS), funded by the U.S. Agency for International Development. Of note is the project timeline and budget: $6 million over four years (2014–17).[29] The extraction of knowledge and technique is fundamental to natural resource management and natural regeneration activities, producing an explosion of conservation missions.

Dryland conservation tends to play out in the interface of soil erosion, between micro-topography, drought, and rainwater. Seeds—slow and iterant—do not associate with the terms of soil conservation. In each *zaï* pit, dormant seeds mingle with seeds nestled in the farmer's fertilizer. Once the first rootlets break the seed coat, they will pursue the water table, inviting hosts of mycorrhizal assistance in order to find the path of least resistance. If the environment changes rapidly, if it is concretized, sprayed with chemicals, infused by sewage, burned, felled, or otherwise disturbed, fitter individual seeds will retain a memory of the changes and adjust accordingly.[30] This is what Margulis calls perception, as she attributes an awareness of the outside world to all *living* beings. The dissipation of each micro-adjustment in *zaï* ecology offers a refreshing approach to expanding how landscape is made and remade over time. Seeds do not associate with the timescales attached to donor societies and the mechanization of the soil that informs project details. Dryland development models that proliferate attention to practices such as *zaï* continue to detach food supply from fuel resources. The clever division between herding, or manure capture, and trees or *zaï* pits is one such example. Prior to colonial rule, trees were commonly spontaneous, owned by farmers or shared by village commons.[31] Practical knowledge confirms that woody plants are integral to crop farming but a hindrance to drylands because they draw on precious water resources. Slowly, practical knowledge is replaced by yield metrics, aided by species lists and classification maps.

The history of tree planting in the sub-Sahel suggests that afforestation in Africa emerged with the spread of the term *desertification,* as administrative influence embossed the landscape by creating a strict division between forestry and agriculture. This was achieved by a mix of botanical and cartographic authority that endures to this day. Consider the "indicative list of the vegetal species adapted to ecological zones," a table of thirty-seven species selected for a supracontinental project (see Figure 2). Significantly, this is the official list of the project, published with a short list of citations. The citations shed light on the sources embedded in each specialization: a mix of florific data by colonizing nations, a reference to the careful work of Aubréville, and a UNESCO "vegetation" map of Africa from 1986. The "Vegetation Map of Africa" is a compendium of various existing map sources for different regions and countries, which were integrated and synthesized by the AETFAT (Association pour l'Etude Taxonomique de la Flore De l'Afrique Tropicale) committee, headed by Dr. F. White of Oxford University.[32] It is a particularly useful map for delaminating how the intellectual work of specialists becomes spatially relevant to the GGW project.[33] While a complete study on the application of the map is tempting, for now consider that White's map delineates bands across the sub-Sahel, as stripes serve as the longitudinal trace of project extents. The bands help unify the project by relating bands to the cartographic legend. However, the legend was never intended as a color-by-number-type experiment. Rather, it was developed to provide UNESCO with a standard from which to attend to physiognomy and floristic composition, not climate, as stated by White. The specificity of extraction lies somewhere between how the sandy colors of semidesert and the yellow fills of grassland and shrubby grassland are simplified to project extents.

The map is well cited, reproduced, and reformatted with green tones in order to reiterate concern. A simple Google search will reveal its popularity and wide acceptance. It will also reveal the simplified drawing and not the original one. The original yellow tint of White's grassland outlines is subsequently replaced with alternating green tones to make the work more legible. Green is the color of choice in afforestation campaigns, although the map was developed to promote closer attention to the region following the cascade of severe droughts in the

No.	Species	Distribution	Ecology	Uses
1	*Acacia raddiana* Savi / *tortilis* (Forsk.) Hayne	Senegal, Mali, Niger, Chad, Sudan, Eritrea	- 50 to 1,000 mm - Sandy (anchored dunes), ferruginous, tropical soils, sandy silt, lateritic talus	Energy wood, lumber, fodder, medication
2	*Boscia senegalensis* (Pers.) Lam. ex Poir.	Senegal, Mali, Burkina, Niger, Chad, Sudan	- 50 to 1,000 mm - Sandy-argillous (consolidated dunes), argillous, rocky, stony soils	Food, fodder, medication
3	*Acacia senegal* L. (Willd.)	Senegal, Mali, Niger, Nigeria, Chad, Sudan, Eritrea	- 100 to 800 mm - Arenaceous-loamy (fossil dunes), lightly loamy (depressions), lithosols	Gum (food, medication, cosmetics), fodder, energy wood, lumber
4	*Acacia nilotica* (L.) Willd. ex Del. var. *adansonii* and var. *tomentosa* / *scorpioides* L. (A. Chev.) var. *adstringens* Schum	Senegal, Mali, Niger, Nigeria, Cameroon, Sudan	- 100 to 1,000 mm - Deep, arenaceous-loamy (fossil dunes), argillous soils, river shores	Lumber, energy wood, fodder, food, medication, gum, tannins
5	*Acacia mellifera* Benth.	Nigeria, Chad, Sudan, Eritrea	- 250 to 500 mm - Argillous soils	Fodder, lumber, energy wood
6	*Cadaba farinosa* Forsk.	Senegal, Mali, Burkina, Niger, Cameroon, Chad, Sudan	- 250 to 500 mm - Sandy soils (consolidated dunes), rocky, pool shores	Energy wood, food, fodder, medication
7	*Cadaba glandulosa* Forsk.	Burkina, Mali, Niger, Chad, Sudan	- 250 to 500 mm - Stony soils	Fodder
8	*Calotropis procera* (Ait.) Ait.	Senegal, Mali, Burkina, Niger, Chad, Sudan	- 250 to 500 mm - Degraded soils	Medication, fibers, fodder
9	*Capparis decidua* (Forsk.) Edgew.	Senegal, Mali, Burkina, Niger, Chad, Sudan	- 250 to 500 mm - Sandy soils, river shores, pool shores	Medication, fodder
10	*Commiphora quadricincta* Schweinf.	Niger, Nigeria, Chad, Sudan, Eritrea	- 250 to 500 mm - Sandy, argillous, and lateritic soils	
11	*Ficus ingens* (Miq.) Miq.	Senegal, Mali, Niger, Nigeria, Chad	- 250 to 500 mm - Spring rocks	Medication

No.	Species	Distribution	Ecology	Uses
12	*Ficus salicifolia* Vahl	Niger, Chad, Sudan, Eritrea	- 250 to 500 mm - Mountain rocks	
13	*Grewia flavescens* Juss.	Senegal, Mali, Niger	- 250 to 500 mm - Pool shores, sandy, argillous, stony, and lateritic soils	Medication, food, fodder
14	*Grewia tenax* (Forsk.) Fiori	Senegal, Mali, Niger, Sudan	- 250 to 500 mm - Rocky, argillous soils, pool shores	Fodder, food
15	*Grewia villosa* Willd.	Niger	- 250 to 500 mm - Sandy, rocky, stony, hardpan soils, river shores	Fodder, medication, food, lumber
16	*Leptadenia pyrotechnica* (Forsk.) Decne. / *spartium* Wright	Senegal, Mali, Niger, Chad, Sudan	- 250 to 500 mm - Sandy soils, dunes	Medication, fodder, log fender, food
17	*Maerua crassifolia* Forsk.	Senegal, Mali, Sudan	- 250 to 500 m - Sandy soils, dunes	Fodder, lumber, fruits, medication, food
18	*Maerua oblongiflora* A. Rich.	Niger, Chad, Sudan	- 250 to 500 mm - Sandy soils, dunes	
19	*Maerua aethiopica* Oliv.	Niger, Chad, Sudan	- 250 to 500 mm - Sandy soils, dunes	
20	*Salvadora persica* L.	Senegal, Mali, Niger, Chad	- 250 to 500 mm - River, lake, and pool shores	Lumber, kitchen salt, medication
21	*Tamarix aphylla* (L.) Karst.	Sudan	- 250 to 500 mm - Temporary water points	
22	*Tamarix senegalensis* DC. / *gallica* L.	Senegal, Niger	- 250 to 500 mm - Sandy (dunes), salty soils, brackish depressions, river shores	Medication
23	*Ziziphus mauritiana* Lam.	Senegal, Mali, Niger, Burkina, Chad, Cameroon	- 250 to 500 mm - Sandy, rocky soils, river shores	Edible fruits, medication
24	*Acacia laeta* R. Br.	Mali, Burkina, Niger, Nigeria, Chad, Sudan, Eritrea	- 250 to 750 mm - Arenaceous-argillous, rocky and stony soils	Gum, fodder, lumber, energy wood

No.	Species	Distribution	Ecology	Uses
25	*Combretum aculeatum* Vent.	Senegal, Mali, Burkina, Niger, Chad, Sudan, Eritrea	- 250 to 800 mm - Sandy, argillous, stony soils on termites' nest	Energy wood, food, fodder, medication
26	*Commiphora africana* (A. Rich.) Engl.	Senegal, Mali, Burkina, Niger, Nigeria, Cameroon, Chad, Sudan, Eritrea	- 250 to 800 mm - Sandy, argillous, and lateritic soils	Lumber, energy wood, fodder, food, medication, insecticide
27	*Cordia gharaf* (Forsk.) Ehrenb.	Senegal, Mali, Burkina, Niger, Nigeria, Chad, Sudan, Eritrea	- 250 to 800 mm - Rocky soils, river shores	
28	*Grewia bicolor* Juss.	Senegal, Mali, Niger	- 250 to 800 mm - Pool shores, sandy, stony, and lateritic soils	Medication, fodder, food, lumber, energy wood
29	*Maerua angolensis* DC.	Senegal, Mali, Sudan, Eritrea	- 250 to 800 mm - Sandy soils, dunes	Food, lumber, fodder, medication
30	*Ziziphus nummularia* (Burm.) Wight et Arn.	Mali	- 250 to 800 mm - Sandy and rocky soils, river shores	
31	*Acacia seyal* Del. / *stenocarpa* Hochst.	Senegal, Mali, Burkina, Niger, Nigeria, Cameroon, Sudan	- 250 to 1000 mm - Argillous and stony soils, floodable depressions liable to flooding	Fodder, gum, energy wood, lumber, medication
32	*Balanites aegyptiaca* (L.) Del.	Senegal, Mali, Burkina, Niger, Nigeria, Cameroon, Chad, Sudan	- 250 to1000 mm - Great ecological amplitude, sandy, stony, argillous, alluvial soils, pool shores	Energy wood, lumber, fodder, food, medication
33	*Boscia angustifolia* A. Rich.	Senegal, Mali, Burkina, Niger, Nigeria, Chad, Sudan	- 250 to 1000 mm - Rocky, lateritic, and argillous soils, pool shores	Energy wood, lumber, fodder, food, medication
34	*Boscia salicifolia* Oliv.	Niger, Chad, Sudan, Eritrea	- 250 to 1,000 mm - Coarse-textured soils	Energy wood, fodder, food
35	*Acacia ehrenbergiana* Hayne / *flava* (Forsk.) Schwfth.	Niger, Chad, Cameroon, Sudan	- 300 to 400 mm - Sandy and argillous soils	Fodder

No.	Species	Distribution	Ecology	Uses
36	*Acacia hebecladoides* Hams.	Chad, Nigeria, Northern Cameroon	- Silt soils	
37	*Rhus oxyacantha* Cav.	Niger		

Sources:
1. A. Aubréville, *Flore Forestière Soudano-Guinéenne* (Cameroon: A.E.F. ORSTOM, 1950).
2. F. White, *La végétation de l'Afrique: Thesis Accompanying the Vegetation Map of Africa* (Paris: UNESCO, 1986).
3. H.-J. Von Maydell, *Arbres et arbustes du Sahel: Leurs caractéristiques et leurs utilisations* (Weikersheim: Verlag Josef Margraf, 1990).
4. M. Thulin, *Flora of Somalia*, vol. 1. (Kew: Royal Botanical Gardens, Kew, 1993).

FIGURE 2. Indicative list of the vegetal species of the Great Green Wall initiative. Note the absence of *Faidherbia* among the thirty-seven species suggested for afforestation. Afforestation schemes profit from the procedures of *artifact* because plants are rendered lifeless. Figure redrawn after GGWSSI and CEN, "Great Green Wall: Implementation," *Implementation Operational Procedures: The Layout and Indicative List of the GGW Vegetal Species* (2018): 8–9.

sub-Sahel in the 1970s. The map offers a way into rationalizing the band, or wall metaphor offered by the project, because the sub-Sahel as a "band" is another form of extraction aligned with the discovery of the wood-fuel crisis in Africa. The famine and the crisis events emerged at a time when the world was gripped by the energy emergency, following the oil-price shocks of the mid-1970s, another disaster of global consequence. As energy analysts and anthropologists began to pile up evidence about the scale of wood-fuel use in Africa, and the millions suffering famine, both types of crisis were essentially rendered similar through White's cartography.[34] The paired crises sparked the reintegration of forestry and agriculture, albeit from the determinations of policy and rural development. In the 1970s, the model of management that kept agriculture and forestry apart began to erode with the convergence of the wood-fuel crisis and the Sahelian famine, as state forestry control in Africa embraced global aid and the promotion of forest plantations. Because research is expressed as a standoff between famine and drought across the entire sub-Sahel, deliberations promoted aggregate scales and universal knowledge, despite the material and daily struggles to find wood to burn for cooking.[35] The local and the global begin to pile up evidence that settles upon deforestation and desertification as common culprits. Planting trees, or the cultivation of wood-fuel resources, offered an immediate response, positioning afforestation once again as a means to satisfy a variety of funding streams.

The notion that the global energy crisis and local testimony of crop failure are codependent was a maneuver to test a global solution, conflating the wood-fuel crisis with famine. At once, this tactic salvages both the prominence of international aid and the village woodlot schemes organized by regional government forestry departments that took over colonial policy. As evidence accumulates that afforestation is possible and even an enduring, historic practice, global attention shifts to the simplicity of planting trees rather than the complications of ending regimes that profit from violent deforestation, the most obvious solution to degradation. The authority of extracted evidence not only has the capacity to guide the production of facts but enhances a social practice that brings voluntary, nonprofit, and charitable organizations together in the service of foreign donors working through local crisis to offset their own poor performance elsewhere. Across the sub-Sahel, the context is so varied that afforestation is conveniently positioned to counterbalance deforestation. The solution economy advances facts without context.

The GGW accumulates evidence by extracting planting efforts, or cases that do not belong to it, a form of environmental extortion that placates relationships across vastly differentiated contexts. Although extraction is typically related to violent withdrawal and depletion, the evidence of mining "success" stories is an abuse that can only be hidden at global scales. The promotion of green-wall planting disregards the qualities of plant life that Goethe and Arber were so content to defend. Tree-unit accounting rehearses the same pacification as taxonomical records, which also means it ignores any collaboration between humans and plants evidenced in each example. Statistics seem to offer a more passionate relationship at the scale of environmental policy, a means to define the political rather than the biotic. In order to explain the transfer of knowledge that absorbs more than it offers, a closer look at one of the first shelterbelt planting campaigns helps expose the hidden layers of exploitation and extraction.

The Story of a Windbreak Project

The landmass of Niger lies primarily in the Saharan desert, a true desert that inhibits fixed human settlement. As a result, most of the population is clustered along the southern borders of the country, the indeterminate range of grasslands in the sub-Sahelian range. As rainfall increases to the south, semidesert grassland grades into low wooded grassland, as height and density increase accordingly. This dynamic climatic range hosts valleys that support fixed settlement. The word *Majjia* is Hausa for valley, a term that appropriately describes the sloped environment around an ancient lakebed in south-central Niger.[36] Here lies the Majjia Valley windbreak project, which is alternatively called a reforestation scheme, a field-planting initiative, a community forest, and a dune-stabilization project. More commonly the project represents one of the most effective initiatives of organized dryland planting, supported by the vast array of studies, publications, and reports that continue to review the project.[37] This claim is also reinforced by the evidence of remnant shelterbelts that contour the region, some of the oldest traces of afforestation in the Sahel. While the project is used as a replicable model for the GGW, the insistence on replication exposes the conventions that structure organized cultivation in order to maintain control over the traditions of the culture it enters into.

The Majjia project is repeatedly cited as an unsurpassed model of locally supported planting, in turn growing expertise and citation in order to execute similar projects under a variety of different social and environmental conditions. First installed to minimize the localized effect of erosion, the project now lends itself to other crises, including desertification, and more recently it is held up as a "successful" typology of agroforestry.[38] The use of typology provokes an accumulation of units that disregards the enormously variegated contexts under consideration, especially in the mutable and dynamic ranges of grassland within Niger. Majjia is an exemplary case because it was initiated and maintained by the community and was not instigated by global narratives of decline. In Majjia, the success of plant life is hinged on its relationship with human settlement, in much the same way as human life is hinged on its relationship to plant life.

The Majjia Valley is home to a shrub savannah typical of the sub-Sahel, with characteristically low rainfall norms that hover around 450 millimeters annually. During the long dry season, the Harmattan winds actively displace vast amounts of particulate that often manifest as dust storms. These storms have a desiccating effect on soil and can cause significant crop damage. As a result, the forty or so villages in the valley hug the seasonally flooded plains, where lowland soil is moderately more fertile. This ground simply does not get picked up and displaced by the wind as often. Based on ancient settlement patterns, villagers in the Majjia value this slight elevation change and in turn profit from a pastoral landscape. Although trees are an important component of the sloped valley, only a scattering of trees remained by the mid-1970s because tree cover had been lost to an upsurge in fuelwood consumption as villages grew but new trees were not cultivated.[39] The CARE "Majjia Valley" afforestation project was initiated in 1974 but was not catalyzed by the observations of experts or the indices of global institutions. Rather, villagers approached a local forester named Daouda Adamou when soil erosion began to affect basic sustenance.[40] According to a field report by a Forest and Society Fellow of the Institute of Current World Affairs:

> The valley's residents were themselves interested in planting windbreaks, as they were concerned about the wind erosion in the valley. They had great confidence in Daouda and had already experienced the success of the woodlots, in producing wood needed locally for poles.[41]

According to a number of reports including interviews and assessments, Daouda conceived of a plan whereby financing was provided by CARE, and nurseries and technical assistance would be provided by the Nigerian Service of Forests and Fauna, but labor and land would remain firmly the responsibility of villagers. This established a vested interest in the success of the project, engendered by the fact that the idea came from an understanding of local vulnerability.

The undeviating linear extents of each planted row reveal the project as a shelterbelt or windbreak planting, not an agroforestry initiative. The first of these windbreaks was planted in the northern part of the valley around Garadoumé, established as double rows of trees, 4 meters (14 feet) apart, with 100 meters (330 feet) between the double rows. As the project moved south, it included the land adjacent to Bouza and Ayaouane. Many of the rows exceed 1 or 2 kilometers in length. According to a field report from 1985, 314.7 kilometers of tree lines were planted perpendicular to the wind, totaling 121,600 tree units.[42] A report from 1991 expresses that tree lines total approximately 10 percent of the valley floor, partitioned into 460 kilometers of windbreaks across more than 20 kilometers; the age and placement of the rows suggest that the majority of the planting was achieved in the first ten years.[43] The Majjia project statistics are also extracted to argue for trees as wood fuel, creating an inventory of some of the most aged windbreak lines. The outcome found that on average one kilometer of windbreak contains about 110 cubic meters of wood.[44] The conflict between specialist assessments are context enough for the model of extraction through appropriation because planting trees for erosion and felling trees for wood fuel are contradictory practices.

Regardless of statistics, the magnitude of the project is ambitious on a number of levels, especially as it relates to overcoming the insistence on peripheral specialists or global aid. What is particularly significant is the support of both laborers and landowners, motivated by an individual steward with knowledge of tree planting. Yet its "success" is corrupted by increased interest in the project as a replicable formula, which misplaces the agency embedded in traditional farming communities that includes familiarity with the ground, plant life, and the climate.

Although species selection is obscured in early reports, subsequent accounts suggest that nurseries prioritized small thorny Arabic trees such as *Acacia seyal* and *A. Acacia scorpiodes* but emphasized the locally celebrated neem tree or *Azadirichta indica*.[45] Neem is an "introduced" plant to the sub-Sahel but is so well adapted and integrated into daily life that its provenance is less known than its beneficial medicinal and astringent offerings and because the leaves are collected for local culinary customs. As a result, neem is unlikely to be cut down for fuel, as its contributions are respected. The Majjia Valley project also introduced the formidable mesquite trees *Prosopis chilensis* and *P. juliflora,* which produce high pod yield and propagate or cross-pollinate freely. As a result, windbreaks are also locally known to "spread spontaneously."[46] As mesquite trees flourish, they outcompete smaller, slower shrubs but continue to provide windbreak protection. Biological nitrogen fixation is another benefit of the selected species, as first succession plants produce nitrogen compounds to help the roots develop.

The cycling is a form of fertilization as the root nodules and the rhizobium associate in the rhizosphere.

Eucalyptus camaldulensis is also a plant popularized by the project to relieve erosion. Also called red-river gum, this plant displays great strength and durability to a variety of conditions, as it prioritizes shallow horizontal roots that sprout readily and can shoot up stalks that even reach three meters per year.[47] This makes *E. camaldulensis* an ideal candidate for fuelwood. The woody plants that were selected across the decade of planting are advantageous for both their suitability to the climate and their symbiotic fixation, but they are most suitable for the variety of local benefits that are differentiated enough not to outcompete crops.

The climate in the sub-Sahel is marked by the slow and persistent registration of seasonal shifts that manifest differently across its varied profile, influencing crop selection. One of the most physical indicators of this shift occurs in proximity to or from a water source. As such, there are marked differences between elevation and the water table that support or hinder development projects such as shelterbelt forestation. The Majjia Valley project is located along the northern fringes of the Niger River basin, located in the Sokoto watershed. Alluvial soils and a relatively high water table characterize the semiarid riverine network that weaves through the valley. The landscape formerly supported groundnuts and cotton, remnant introductions from colonial decree. Significantly, ongoing climate shifts are registered through environmental cues that manifest physically in plants, as species composition shifts. The registration of climate is evidenced in both the wild and cultivated plants, as cultivation of hardier and more simplified crops such as millet and sorghum have now replaced groundnuts and cotton.

As an organized project, the story of the Majjia Valley windbreak advances a potential shift in practice between human and plant life. It locates knowledge on the ground, between valley formations and low points near the water table. The design acknowledges the extant landscape, as limited water supplies are valued in relation to the uptake of different root systems. Soil relations are prioritized over the subsequent success of food production and settlement in the region. The acknowledgment of limits fortifies the "success" of the Majjia Valley project and confirms that a windbreak resists scalar abuses because tree planting cannot reiterate across drylands with the same outcomes. While the typological evidence of linear windbreaks is perceptible, and the species selection can be simulated in pilot projects, the mingled attributes of the landscape cannot be replicated because the windbreak is living in context, and calibrates to the scale of the village. In drylands, the benefit of planting a tree is found in establishing a relationship between human and plant life, and relationships resist typology. Perhaps accomplishment can be valued in the networks of communication that lie outside the tendency to "scale up."

The Majjia project remains an anomaly of "success" because it progressed with the provisions of a local resident and the support of an active community and not because of expert species selection or botanical specializations. The project went unnoticed until a decade of growth and stewardship drew attention to its "success" reported from a number of personal accounts that absorbed its significance. For instance, the failure of reforestation schemes is commonly blamed on the lack of general awareness about established forestation techniques. A windbreak will not function if it is felled for fuelwood, which is often the reason for failure valuation, as blame is cast on the lack of community knowledge, or lack of expert knowledge. Yet, when Hausa households were interviewed, responses suggested that the windbreaks were not felled for fuelwood due to respect. Respect is not feature of typology; it is a feature of relationships. It turns out that policy change does not necessarily transform knowledge into spatial evidence. This is a common supposition of reforestation schemes that blame the lack of general awareness about *established* forestation techniques. But CARE was not responsible for transmitting respect for windbreaks, just as it turns out that policy change does not necessarily transform knowledge into spatial evidence.

In the sub-Sahel, experience is mired in colonial inheritances that continue to corrupt decision-making. Before the advent of the Great Green Wall of Africa, tree planting and afforestation left out the transmission of nonmaterial consequence: "The social researchers found that most local residents do not believe that they own the trees—most think that the trees belong to the local forester to the government."[48] The interviews reveal the inheritance of ownership, once the domain of colonial forestry, inspiring CARE to support a woodcut that distributed wood from the shelterbelt trees among the villages. While the effects of policy recommendations that are based on colonial sources are well reviewed, the spatial repercussions continue to perpetuate an inaccurate narrative of production and management. Remarkably, it was not until 2004 that state forest code in Niger formally recognized community rights within forest reserves and established customary rights to use forest resources located in areas held by local agroforestry initiatives.[49] The Majjia Valley project reveals that the motivation from a village merged with experiences and contextual circumstances in ways that resist replicable models. Nevertheless, the project is absorbed by the GGW committees to provide evidence to the humanitarian elite.

3

Genus *Faidherbia*

FAIDHERBIA ALBIDA IS A TAXONOMIC CONUNDRUM. It confounds specialists because it resists universalizing categories. Early discoveries of the plant located it within the genus *Acacia,* since it tends to mature along with *Acacia sieberiana, A. etbaica, A. tortilis,* and *A. seyal.*[1] The story of *Faidherbia* is found in its unique defiance of category, a behavior of resistance that isolates it from other *Acacia* in the genus. Because of its defiance, *Faidherbia* is labeled a monospecific genus, which means that it finds itself in a taxonomic category that contains only one species, an isolated "species" category that allows *F. albida* to linger until more plants can be added to its kind. Monospecificity reminds botanists that there are no other extant correlated plants or that we have yet to find other *Faidherbia.*[2] This commotion in relationship status advances the inclusion of behavior when considering a description of plant life, providing cues to the breakdown of industrial, inventorying organizations of the plant as artifact.

The commotion to name and distinguish plants highlights why studying behavior informs one of the most basic parameters of afforestation: species selection. *Faidherbia* is a highly sought-after tree for afforestation and agroforestry initiatives in the sub-Sahel. It is rationalized for its aboveground qualities by reference to its external form. It appears through the priorities of species selection on afforestation lists, absent any indication of behavior, relations in the rhizosphere, or associations with the climate. As we shall see, it also overlooks the difficulty of growing *Faidherbia* as nursery supply. Selection highlights the discrepancy between growing trees and planting trees, as the plants of afforestation tend to be traded by the currency of the artifact. Once planted, or transplanted from a plastic pot, *Faidherbia* grows according to its own rules. *Faidherbia* defies authority beyond the question of epithet.

Disagreement over description began in 1829, with *Faidherbia*'s first unearthing and identification. Crucially, the debates between *Acacieae* orders exist to decide if the plant is "productive," a process of elimination or exploitation. Either way, conduct is muted and parts are cataloged. Even as the seeds of *Faidherbia* are scored, soaked, engineered, and mechanically planted into the sandy soils of the Sahel, the indexical strata of taxonomic debates remain "unresolved."

> Tribe *Acacieae* is widely attributed to Bentham (1842), e.g., Vassal (1981) and Maslin et al. (2001), but Reveal (1997) gives Dumortier (1829) as the first place of publication of the tribe and this is confirmed by Brummitt (pers. comm., 2004). The genus *Faidherbia* A.Chev. was included in the *Acacieae* by Vassal (1981), and is still retained as part of the tribe by Maslin et al. (2003). The tribal position of *Faidherbia* remains equivocal, although Lewis & Rico Arce (this volume) place the genus in tribe *Ingeae* following Polhill (1994) and Luckow et al. (2003) rendering the *Acacieae* monogeneric, with the single genus *Acacia*. The taxonomic status of *Acacia* and its relationship to other mimosoid genera is, however, as yet unresolved.[3]

As the plant collapses into language, species and tribes are attributed to the findings of plant collectors that lay claim to origin status, elucidating a breakdown of the plant into units. Through publication and reference, science is telling stories of heroic speculation, turning "unresolved" matters into appraisal, transforming everything into an opportunity. This story—of turning science into politics—is well known and does little to redistribute agency to the plant itself. The question is whether the inclusion of behavior might inform the specific actions of environmentalism, collaborating with *Faidherbia* to counter afforestation.

In its biological, living state, *Faidherbia* survives in dryland soils without rainfall recharge because of lasting belowground relationships and the coevolution between buried resources and climate extremes. It is a coveted tree of afforestation because it withstands extremely long drought periods, flourishing without rainfall. *Faidherbia* is both fodder and fuel, as it is associated with prodigious growth in pastoral and nomadic traditions. In other words, it adapts and prospers whether humans settle or roam. *Faidherbia* is a plant that produces its own requirements, its own actors, and its own sustenance. But it resists one of the most fundamental procedures of artifact: it opposes vegetative propagation and nursery standards. By unveiling the unique qualities specific to monospecific *Faidherbia,* opportunity shifts, reversing the interests of human practices alone.

Tree-planting campaigns in drylands exemplify the global tendency to exploit an agenda or a crisis external to regional dynamics. Unfortunately, Africa is no stranger to this abuse. And curiously, each crisis is managed by a reliance on the image of a tree. In its formal state, *Faidherbia* emerges as "tree," rendered by international agencies as an agroforestry product or an afforestation technology, while making its way more recently into conservation schemes and ecosystem service calculations. Either way, tree planting regardless of context is justified as a brand of crisis management. The tree prevails, even as the hazards of each crisis range drastically from imminent overexploitation of resources to trends in agricultural technology—from the green revolution to the Anthropocene.

Paying attention to the temporal conduct of *Faidherbia* shatters the assumptions of artifact. The behavior expressed between plant life, human life, and the climate suggests a unique description of the African Sahel, between the climate and the ground, with *Faidherbia* as an intermediary: a fine collaboration between species that fuses microorganisms belowground with the activates of seasonal change aboveground. In particular, the expression of behavior is located in three main reproductive adaptations that amalgamate in the lively world of *Faidherbia*: the formation of extremely deep penetrating roots, the seductive rhythms of radical phenology, and the careful maneuvers with other organisms as diverse as microscopic bacteria and clumsy grazing ungulates. Overall, each decisive behavior reinforces *Faidherbia* beyond categories, an exceptional living assemblage beyond rank.

THE ROOT–RHIZOSPHERE BOUNDARY

Faidherbia is not a showy plant. It does not display formal characteristics that make it appealing to humans. It does not flower brightly, fruit intensely, or achieve an external silhouette of great splendor. *Faidherbia* is also known as the ana tree and the apple-ring tree, a survivor that invests all its energy and resources into establishing itself for extended durations—for the long haul. It does so by first establishing a taproot that blurs its associations in the rhizosphere, not unlike the difficulty of drawing a line or establishing an operational definition. Experiments collapse when *Faidherbia* is made the subject of growth rate analyses, solution measures, or plantation infrastructure, as the seedlings seemingly self-select.[4] How does plant life self-select?

The root that arises directly from the seed coat is called the primary root, formed by the elongation of the radicle. So, taproots are primary roots that invest in growing straight downward from the stem. A root is just an underground stem, driven by a different tropism, the cues taken in from the outside world. It is worth noting the distinction between roots and rhizomes; rhizomes are horizontal and capable of producing both shoots and roots. The main difference is that rhizomes function more like stems, although they swell underground. Conceiving of the living, active movement of the root and rhizome as a system and an extension enlivens the space below the surface of the soil and propels a reading beyond parts. The complexity of root penetration depends on a number of factors, but elongation is the main achievement of the primary root. *Faidherbia* determines its status when it sends out the primary root, the first "organ" to emerge during germination once a seed has nestled itself into just the right position. This emergence is primary but endures as a taproot.

Taproots are characteristic of plants with long individual lives. The taproot dives quickly to secure a steady supply of water for the growing plant, as it pierces the soil in direct correlation to the depth of the water table. Plants that root deeply to reach the water tables are called phreatophytes, which as a result enables these species to produce new leaves at the beginning of the dry season. *Faidherbia* only finds respite from its powerful downward advance once the root tip is close to this source, spending energy absorbing water rather than dividing cells.

Plants with taproots invest in securing both support and conduction before sending leaf organs out into the atmosphere. *Faidherbia* taproots are not individual organs (like a carrot) but splinter into a mesh of activity and rhizoidal outgrowth that is governed by resource acquisition, growing in tandem with signals in the rhizosphere. This root–rhizosphere boundary is a world of cohabitants presenting another confusion for specialists with their heads bent to the canopy.

It is difficult to draw a distinct line between roots and their co-habitants in the soil, because sloughed-off root cap cells maintain metabolic activity while detached from the plant, mycorrhizas provide a continuum from plant to fungal tissue, and cell walls leak soluble carbohydrates. Therefore, no simple operational definition of the root–rhizosphere boundary exists that allow unequivocal measurements of the respective respirational activities.[5]

As the root structure matures, it sends out a vast network of smaller roots called rootlets that further complicate the root–rhizosphere boundary. Each root, regardless of its size, makes sense of the substrate by communicating to secondary and tertiary roots, such that their roles become defined between surface feeders that grow laterally and deep divers that sink to secure other levels of potential nutrition. There is no scientific monitoring that can account for this sprawl. Measuring uptake in roots is rarely achieved in situ, just as any attempt to qualify the relative contribution of given soil layers to the water table is challenging. The underground world of plants remains a topic of speculation, for any meaningful study displaces and disrupts, another particularity unique to the study of behavior.

Consider a description of dry-season plant growth within a managed park in Sudan. The authors of one study cite agroforestry as the primary keyword, signifying that their achievement will be endorsed to boost planting units. In trying to delimit metrics of "efficiency," this experiment slips into what can only be called a theory of the root system.

The trees were trapping ground water at large depth in the soil. The root system was complex, and colonized several soil compartments, with different water dynamics. Trees were likely to exploit all compartments at different periods of the year.[6]

The supposition that these well-monitored trees were "likely" to exploit the horizons of the soil is a perfectly reasonable evaluation, one that would not immediately seem to necessitate the trials of scientific proof. The authors cite a variety of metrics and constraints associated with their "experimental stand," fully acknowledging the deficits in their study by including statements such as "little is known" and "there is strong evidence." Articles such as this demonstrate how *Faidherbia* was cast as the hero of agroforestry, designating *Faidherbia* as the species of choice across Africa drylands, from intercropping to afforestation. Despite the clear ambition to "improve" plantations, the authors of the experiment reveal root behavior, the comportment of the plant that might not be exploitable but can certainly advance a study of the plant—for its own sake.

Rather than appreciate the surprises and unknowns of behavior, specialization reveals the solution–conclusion conundrum that justifies afforestation. The study of plantation improvement conveniently resonates among the community of governmental, nongovernmental, and scientific organizations that are interested in how *Faidherbia* consumes water from depths far below the soil horizon of crops. But the uptake cannot be measured. In drylands, *Faidherbia* enables annual crops to flourish because it does not limit water supply. Or put more bluntly, it will not compete with irrigation systems. The study is not a confirmation of root behavior; it is a confirmation of irrigation technology.

While *Faidherbia* inspires biotechnology and provokes policy, the primary root or taproot continues to penetrate the fine composure of dry, dusty soil, advancing through signals perceived by the cells of its root cap. A root cap develops from the tip as a group of meristematic cells that keep the plant growing. Meristematic tissues have the capacity to divide and are responsible for the primary and secondary growth of plants, in contrast to permanent tissues, which are produced by meristematic cells but have lost the capacity of division. Typically, root cells divide at the tip or apex of a plant organism, but not all cells of a plant are capable of repeat division. In other words, meristematic activity occurs in growth patches, modules, and tips among juvenile cells, below the surface of an existing root or shoot. At the same time, lateral roots—any branch of another root—provide the physical support for the aboveground plant to grow while transferring nutrients and water back to the taproot, sprouting horizontally from the taproot and increasing the surface area of the system in order to make more contact with the soil. The root apex or primordium is a group of cells at the tip of the root that are in this early stage, typically newly emerged or actively growing. So, the root primordium is the principal driver to the establishment or breakdown of a *Faidherbia* seed, as the first signs of life pierce the outside of the waxy seed coat.

The division and enlargement going on behind the root primordium drives the taproot to increase in length and girth as it advances through the soil in search of the water table. This is where the thrust of movement occurs, underground and out of sight. Although numerous secondary or lateral roots proliferate outward until the taproot reaches the water table, the plant conserves its resources until the soil has been fully explored. In the sandy, exhausted soils of the sub-Sahelian range, such concealed depths attest to adaptation as a force.

While excavation methods vary, extracting the root from the rhizosphere is the only method to reveal the power of growth. It is the only way for humans to observe the influence of the environment on the root and vice versa. Mature trees resist excavation not only because it eradicates the life of the plant but because it significantly alters the actual extents of the rhizosphere. Excavating young trees, on the other hand, is simple enough and can be accomplished within a few years. Regrettably, associations beyond the root are equivalently endless in saplings; they are simply easier to disregard. *Faidherbia* taproots are recorded at remarkable depths, including those beyond the water table at 20, 34, and up to 50 meters in some cases, averaging an advance of 5 meters per year.[7] A controlled study in Sudan found that 8-centimeter seedlings produce a taproot of 70 centimeters, while a study in Nigeria excavated a three-year-old sapling, which displayed a taproot in excess of 9.9 meters.[8] Such studies, although not directly funded to support knowledge of root behavior, reveal that in young tap-rooting plants, the root is significantly more advanced than the shoot. Moreover, the root develops in direct relation to its access to water. The liveliness required to sense the water table and adjust accordingly ensures that *Faidherbia* can survive the extreme heat and aridity, independent of rainfall events, as remarkable depth relations inspire agency beyond the artifact of study.

As the roots dive toward the water table, they communicate or chemically announce their presence to existing colonies of soil bacteria called *Rhizobium*. Although *Faidherbia* is a monospecific genus, it is part of a large family of leguminous plants. Legumes form a partnership with rhizobia, a group of microorganisms defined solely by their ability to nodulate leguminous plants.[9] Rhizobia induce root nodules, specialized structures that mutually benefit the bacteria and the plant. In the rhizosphere, rhizobial relationships only occur along moisture gradients: imagine *Rhizobium* suspended in the dry substrate, sluggish and somewhat isolated until a wetting front saturates their world. Water molecules are met with an excitement that allows rhizobial colonies to speed up and make their way to a consortium of young roots. The roots are greeted by *Rhizobium* that can finally get moving because the soil is saturated, engendering the fine relations between plant life, other organisms, and the climate.

Faidherbia stimulates environmental relationships in drylands. Likewise, these relationships are hard to notice, and they require the patience and care of suprapersonal methods. In describing the study of roots, Arber asserts that our knowledge of the root system falls short of the rest of the plant, and that there is no "short cut to this knowledge."[10] The complications of the underground refuse to give certainty. This is one of the reasons that the root boundary is studied through individual organisms like bacteria and environmental inputs like moisture. As we have seen, moisture levels affect the ability of rhizobial relationships to move through the soil and along the root system, so it is also worth noting that rhizobia will only colonize young roots with emerging root hairs.[11] The partnership begins along the root as it develops, as rhizobia seek rest near juvenile cells, the soft parts of the root from which root hairs extend. As rhizobia and young plant cells mingle, they divide and multiply within a plant-derived membrane, such that there is no clear boundary between them. By growing a nodule designed specifically for another organism, plant life welcomes rhizobia into the developing root system, securing a reliable supply of nitrogen that they cannot obtain from the air themselves. In turn, the rhizobia fill the walls of the plant membrane with bacteria that assimilate nitrogen for the plant host.[12] When rootlets are reliably shed by the plant, the fixed nitrogen is released into the soil, nourishing the rhizosphere and increasing fertility. Once reframed, rhizoidal relationships cannot comfortably be called an infection, as the ability of *Rhizobium* to nodulate legumes depends on special circumstances and specific mutual recognition often termed *partner-choice*.[13] Both organisms benefit, although the plant does not need *Rhizobium* to grow. This is why it is called a choice.

Rhizobia fix nitrogen, an advantage to the plant, the bacteria, the soil, and the climate, a reminder that there is a whole world unfolding, competing, and relating in the rhizosphere. Roots are adapted to the microbes that inhabit the soil in much the same way that the bark and canopy develop associations with humans and animals aboveground. Of interest in soils is the expression of how living material alters itself toward inclusion and acceptance. Plant life rises above individual or exceptional status as it expands. Such collaboration in the rhizosphere resists objectification and reveals why there is no simple operational definition of the root–rhizosphere boundary.

RADICAL, REVERSE PHENOLOGY

Phenology is the study of periodic variation. The term refers to the temporal response of biota to the climate; it is the signaling between atmospheric change and plant or animal reactions— from bud burst in the spring to seasonal bird migration and the timing of first frost. In relation to phenology, *Faidherbia* displays a specialized reverse deciduous cycle that compels a great deal of attention to this monospecific genus. The plant decisively sheds its leaves at the beginning of the rainy season, allowing light and moisture to penetrate the ground and support food crops without overshading them. Being leafless in the wet season and leafy in the dry season is a highly sought-after attribute for humans and animals, one that finds no counterpoint in the Sahelian range. Behavior so attuned to climate variation is a remarkable advantage for the cultural scripts that rely on small-scale sustenance farming, enriching the relationship between human and plant life. Radical phenology confirms *Faidherbia* as a mediator between the soil and the atmosphere, a sentinel of drylands.

The phenology of *Faidherbia* confirms the cycles of cohabitation that are caught between the language of remote authority that urges communities to settle and the realities of nomadic networks. Consequently, the Sudano–Sahelian and northern Sudanian zones are inhabited by both settled farmers and seminomadic tribes. In these human networks, the ability to endure a sedentary life and the urge to remain agile compete through the spatial disparity between pastoralism and farming.[14] This tension is articulated in the southern belt of the Sahel, a region where crisis narratives easily take hold, including the threat of extreme drought in relation to settled farming and the consequences of overgrazing in relation to pastoralism. Each hazard is structured by the dispossession of land created by settler colonialism that results in the physical erasure of indigenous practices. The challenges are reproduced within the global crisis complex of humanitarian, media, and nongovernmental organizations, as policymakers argue over the value of farming versus pastoralism, fixed or mobile patterns. On the ground, the reality of dispossession produced by colonial tactics is multifarious: some people choose to settle in one location and others choose to migrate and remain mobile, while others are still forced in one direction or another. But this is not the issue at hand. Rather, the significance of shifting human settlement patterns is relevant only insofar as they are absorbed by and communicated to plants.[15] Of note is how the same competition dynamics between settlement and mobility are reflected in plant life: for instance, the shifting habit of tall woody plants that intergrade with dry grasslands, and the fixed rows of annuals that describe plantations. As we shall see, plants such as *Faidherbia* evolve to exploit human patterning, using fixed settlement as a means to further their own cycles of reproduction.

Plant life actually advanced with dryness, as the most adaptable species seek advantage of these ever-expanding regions. Botanist-scholar Joseph Armstrong reminds us that plant evolution must be considered across billions of years when considering the mutability of dryness: "The main idea here is that increasing aridity produces new habitats and the very adaptable flowering plants took advantage of these areas. . . . And our familiar groups of animals evolved in reaction."[16] This claim is significant for two reasons: First, it reminds us that aridity migrates along earthly timescales. What is dry now might very well be wet in the future. The second point is that flowering plants dominate the biotic world, from grasses to trees, including *Faidherbia*. In fact, all the plants on afforestation lists are flowering plants or Angiosperms, a surprise because often their small flowers are inconspicuous to humans. Plants that spread across territories with distinct wet and dry seasons are responding to dryness, evolving radical behaviors such as swelling to increase water storage, producing millions of microscopic leaves, and developing significantly more biomass underground than above. For plant life, growing smaller is not considered a shortfall, as evolution is not unilateral. Adaptation to hyperaridity progressed slowly across billions of years, engendering an intelligence that continues to help flowering plants expand in global drylands. In this way, hyperaridity and plant life are not mutually exclusive; rather, climatic and biotic processes collaborate. And they do so across vastly more extended timescales than humans. Aridity and other extremes of the climate are only limited for humans, who portray the condition as an obstacle. Periodic dryness may produce excitement and anticipation for a plant imagining the prospect of unencumbered spread. For instance, "extremophile" is a term given to plants that live happily in conditions that are only physically extreme for humans. Plant life is already familiar with rapid change and extremes of the climate.

As *Faidherbia* trees calibrate to difference, they actually produce the sociospatial patterns of pastoral and agricultural subsistence, quietly acknowledging the requisite of both systems. Effecting and restructuring human settlement patterns does not require the plant to be mobile. Reverse phenology reveals that when *Faidherbia* is leafy, its shade promotes cattle grazing, as herds pattern their movement in the gaps between individual trees. They do this in order to profit from shade. Once these animal movements are recorded and recognized by the plant, the dark yellow-brown pod fruits of *Faidherbia* ripen and drop, offering the hungry grazers a chance to feast on moist legumes, in the shade of a full canopy. At the end of the dry season, the leaves begin to drop slowly. The last pods hold on, as leaves are shed, such that the cattle are lured into staying a little longer, despite a leafless canopy. The last pods and leaves are finally released in concert with the rainy season. Imagine the plant calibrating its drop to the length of daylight hours left in the weeks before the rain starts. If the rainy season is late, *Faidherbia* waits. The plant perceives subtle atmospheric shifts and identifies a corresponding response by coordinating a chemical signal. In this way, *Faidherbia* is both a remote and a direct agent of change.

When all the leaves finally drop, a silent but meaningful message is sent to humans that the rainy season is coming. In turn, humans move cattle into further pastures and begin to cultivate annual crops among the leafless *Faidherbia* trees. The trees are now starkly silhouetted along the horizon of other plants that are taking advantage of the rains through the emergence of leafy organs, greedily absorbing the new moisture of the season and depriving the ground under their canopy. Sowing annual crops under leafy trees is an obvious disadvantage. But *Faidherbia* is leafless in the rainy season, a radical, reverse phenology.

From the perspective of the plant, cultivation and pastoralism represent a seemingly free supply of ungulates, of consumers for their seed. Reproductive success in such a harsh and patchy terrestrial environment requires a dispersal mechanism that can be relied upon not to create too much competition. Across drylands, this interface is associated with cycles of drought that regulate practices. Cultivated land is the most prominent and lasting contact between humans and plants, so any organism that associates well with the culture of plantations will prosper.[17] *Faidherbia* taps into, or evolves within, the mobile paths created by grazing cattle. Cattle wander along unpredictable vectors. Plants that drop seed in their own, more immediate environment would form colonies that seek reserves, an adaptation destined to fail in arid biomes with such low water supplies. Clumping behavior is more noticeable in wet, humid biomes where thickets develop. But plants that clump create too much of a struggle for resources in drylands. Characteristically, plant life tends to take on a distinctly scattered pattern in the sub-Sahelian zone. Widely spaced-out plants extend wide networks belowground to avoid drought stress, as spacing is revealed along limits that correspond to inputs such as available water supply, wind depressions, and soil texture.[18] According to Armstrong, shifting to an immobile lifestyle within drought cycles actually allows plant life to stay in a favorable patch. Moreover, the spacing of plants is not an arbitrary threshold; it suggests a shared economy between species.

AGENTS OF DISPERSAL

In *Plant Sensing and Communication,* Richard Karban explains that drought stress is one of the environmental cues that some plants "remember." Remembering is a controversial reaction in relation to behavior but corresponds biologically to elevated levels of calcium in the ions that help stabilize the permeability of cell membranes.[19] Calcium ions are particularly important to developing fruits, in this case the fleshy part of the leguminous seedpod shed by *Faidherbia.* Remembering ensures spread and survival. Karban also highlights that plants exposed to frequent drought cycles absorb the experience and respond by protecting their kind, demonstrating a close-knit survival based on chemical communication. Here again, dryness can play an important role in elucidating the central role of plant life in forming landscapes. Thus, the plant recruits local externalities (cattle, humans) in order to survive and endure, calibrating response through memory. To the plant, animals are simply a convenient agent of dispersal. The response and memory inform evolution, advanced by radical phenology.

Since *Faidherbia* cannot move about to score mates or catch resources, the evolution of its reverse-deciduous phenology offers a unique swap in return for a diffusion of resources. Plants are primarily concerned with survival through reproduction and evolve reproductive powers in excess of their need. These powers are what we call pollen, fruits, and seeds. In the Sahelian range, herds actively search for and digest the fallen pods of *Faidherbia,* passing any of the nonconsumed seeds back into the environment through feces.[20] Most of these small seeds do not persist, do not make it through the digestive process, or simply do not receive enough light and water to germinate, which is why the plant supplies so many. Under favorable conditions, the seeds that do survive are carried across otherwise impenetrable territory (for the plant), since cattle pathways are pushed further afield as row cropping replaces grazing under the *Faidherbia* trees. In other words, the cattle are deliberately moved out of the way by herding communities as the canopy leaves are dropped; the plant signals the human to take advantage of the leafless canopy for growing food crops. In so doing, the trees ensure that their offspring (seeds) will have the best chance of survival. The plant recognizes the advantage of keeping up the cycle of reproduction that moves the cattle away. Perhaps the phenology of *Faidherbia* evolves to exploit human rhythms, recognizing that in order to secure its rapid spread, the cattle would have to move on.

The seeds of *Faidherbia* are lodged in indehiscent pods, a term derived from the Latin *dehiscentia,* meaning open, or split. Indehiscent pods do not split at maturity in a predefined manner; rather, they rely on predation or decomposition to release the seed from the pod. *Faidherbia* seeds are only discharged after an ungulate has grazed and digested them prior to the rainy season. In particular, flowering plants are prominent in our time period because they present olfactory and visual displays that attract humans and animals through rewards. This reward is what we call food. In return, the plant persists. The field of evolutionary biology acknowledges that these incentives are not a sign of the plant being nice to the animal, or unknowingly keeping the human species fed. Rather, such biological interactions are an exploitative means for the plant to reproduce, because rewarding better pollinators and effective dispersal agents results in more of their offspring. The benefit seems to be directed toward the animal in human terms but is actually a means for the plant to train the most effective agents to have more offspring carrying the genes of their successful parents, securing a cycle that keeps the plant in control.[21] The way in which humans depend on these same interspecies and biological interactions cannot be left out of such a basic evolutionary concept. Plants have coevolved with humans and other animals such that it is impossible to characterize human life without plant life and vice versa.

Once *Faidherbia* has signaled humans to move the cattle, seeds begin to germinate further afield, in an attempt to reorganize the land, since a successful *Faidherbia* crop will attract settlers. Seeds begin this process of reordering their immediate space by prioritizing the establishment of deep taproots. The payoff of germination through plant–animal interaction produces the next lineage of *Faidherbia,* as another generation of offspring adapts to the climate and transforms the landscape. The close link between the phenology of *Faidherbia* and seed dispersal tactics suggests that plant life and the dynamic climate are an environmental force, in which adaptability is key to survival and boundaries between plants and soils, or trees and grasslands, are in constant flux.

Imagine the number of seeds that lay buried under the surface of the earth. They are utterly discreet to the material layers that veil the surface, or the boots and hooves that trample the thin shallows of their habitat. Some are dormant entirely; others are waiting to receive the correct signals in order to flourish; while others are still investigating the substrate with their root organs, deciding if the conditions are ripe for success. *Faidherbia* seeds are actively sensing the proximity of other seeds and the local water table, as they invest all their developmental energy in designing a taproot that can withstand a decade of aridity before beginning to produce enough seed to spread itself further. This means it can take many years for *Faidherbia* to be useful to humans.[22] Most plants are appreciated for their canopy, as description of the visible branching systems can be counted and identified in the early stages of development. Yet detailed investigations of the branching patterns of established tree root systems are very rare. The intelligence of *Faidherbia* is embedded in the emergence of its taproot from seed, which can be honored precisely because it does not sustain prediction and acts independently to the design of planting typology and the standards of the nursery industry regardless of whether or not it appears on planting specifications.

Afforestation schemes always include *Faidherbia* on species selection lists. It is highly sought-after for its reverse deciduous cycle, and it is the fastest-growing indigenous savannah tree in Africa. As a result, efficacious nursery propagation is essential to sustain projects of afforestation. Although *Faidherbia* is an effective tool, it does not respond well to the conventional methods of vegetative propagation that large-scale planting necessitates. The enigma is that *Faidherbia* does not impede heavily laid irrigation lines in landscapes of cultivation that are economically approved through water budgets. Since *Faidherbia* taproots seek water from deep soil horizons, this plant does not compete with the capital and technological investment of irrigated cropping systems that litter the surface. Its behavior is precisely why *Faidherbia* is a great development tool but a nuisance to cultivate. At the same time as this taproot is celebrated for not interrupting drip line, it is contested because it defies nursery standards. The same cells that pierce rock, split clay, and weave toward cavities of nutrient supply in concert with soil fauna and microbial communities will not be undone by plastic pots, glass cylinders, and synthetic mediums provided superficially by humankind.

Micropropagation is biotechnology, an effective way to select clones with good potential for taproot growth.[23] Roots are grown in-vitro, a process called micropropagation, whereby seeds are raised vertically and photosynthesized by a fluorescent tube. Selecting plants with rapid elongation rates is an ambition of determined agronomists whose projects emboss food grain commodities into the dry grasslands of the Sahel, blending a holistic ambition to "feed" and "amend" humans while functioning as a determined developer. The advantage of micropropagation is that it produces many plants that are clones of each other; therefore, such clones can be bred for their disease resistance, viable seed output, or, in this case, taproot vigor.[24]

The determination of the taproot is extremely difficult to measure in breeding trials, yet the genetic disposition of its resolve to dive deep is the aim of many experiments. From the point of view of the root, these trials unfold absent bacterial relationships to confirm overall physical structure and strengthen aboveground vigor. Technology obscures the copious need for biotic inputs, the mineral fertilizers and networks of inoculated mycorrhizae.

Taproots such as the ones that characterize *Faidherbia* enlarge and seek sustenance from their belowground habitat. Despite this expansive geography, roots are not particularly good at actual absorption. Each micro-movement proliferates cells and reaches out to form relationships in the interest of absorbing nutrients and moisture. The ability to extend beyond the "boundary" of the root in order to engage with absorption transpires through root hairs. The extension of each hairy rootlet attracts necessary bacteria, fungi, and other microbes into a mutually beneficial relationship with the rest of the plant.[25] Particular to this relationship are the mycorrhizal fungi that greatly enlarge the available soil volume for absorption through extensively fibrous networks.[26] Mycorrhizal fungi are also microorganisms (like rhizoids) that live concealed in the substrate and support plant communication. Plant–mycorrhizal relationships, unlike the relations between rhizoids and legumes, are highly conditional because mycorrhizae are generalists, capable of associating with many different plant species.

Tree-planting procedures in the Sahelian context do not examine mycorrhizae associations across drylands because trees are imported or grown ex situ. Rather, beneficial mycorrhizae are introduced and inoculated into nursery soils or injected into agroforestry plots to help improve the "well-being" of the root system. The indications this procedure reveals offer little about the collaborative force of the topic, including multispecies associations, the spatial distribution of the rhizosphere, or that plant life makes decisions; rather, each procedure reiterates the dominant narrative of colonialization in which "the poor soils of Africa" can be fixed or improved technologically. Such credentials are usually placed in the introductory remarks as a foundation for a conservationist's report and include such ambitions as "reversing" soil fertility decline, or "amending" highly weathered surface horizons, only to recount that organic inputs are scarce and mineral fertilizers are often out of reach for subsistence farmers.[27] There is no mention of the reasons for such decline, or that decline in this case may just be the consequence of aggrandizing methods that insist upon yield statistics, industrial cropping, and calculating harvest. The deteriorated relationship between humans and other organisms is revealed in the variety of "fixes" that are not only territorial but appear in the micro-considerations of project making, evidenced with mycorrhizae.

Because most soils contain some mycorrhizal propagules already, competition between indigenous strains and introduced strains is a key factor in the response to inoculation. Very little is known of the competitive ability of strains because they are so hard to identify since they cannot be grown in culture. At the moment, competitive strains are selected empirically on the basis of plant growth response to inoculation in non-sterile soil. If the plant responds, the strain is considered more competitive than the native strains in forming mycorrhizae.[28]

In other words, while afforestation projects now insist on indigenous plant material, speculation in the content of nursery soil relies on introduced strains of mycorrhizae that inevitably find their way into indigenous soil. Plant roots engender a community of relations in the rhizosphere and suffer from isolation metrics that treat the soil as a lifeless medium and the plant through its aboveground status.

As the plant develops in the nursery pot, it does so inoculated with foreign propagules and without environmental cues. Once the sapling emerges, it is transplanted into drylands absent a deep enough taproot to actually engage the water table. Thus, the same unworkable intellectual histories that we are confronting in other parts of biological sciences are rehearsed. This includes the deliberations concerning native and nonnative plants, which is a history embedded in early colonial exploits and introductions. Transplanting *Faidherbia* is just another form of extraction that assumes dominance over plant life. In the twenty-first century, it is surprising that these operations are considered "environmental." That mycorrhizal introductions are not a matter of more concern is equally surprising, since the consequences of large-scale ecological interruption are precedented. For instance, eucalyptus was introduced to Algeria from Australia in 1854, at a time when its provenance and properties were unknown.[29] It was planted by the thousands and quickly naturalized into millions of individuals, as it soaked up water from each horizon of the soil, depleting local wells and springs while shading out plants that had adapted to the Sahelian climate. Eucalyptus is now emblematic of exotic plant invasions and the character of declensionist environmental narratives.[30] This assemblage of relations between the invisible roots and the inoculated soil not only repeats the protocols of previous ecological oversights; it does little to resist the temptation to abuse living organisms that are too small to observe, too concealed to infiltrate, or too arduous to identify.

Faidherbia is an ideal dryland partner; it succeeds on its own terms and is indifferent to nursery practices. Failures in nurseries are blamed on its taproot, on its resistance to being potted. It resists the pressure of irrigation and plastic pots and collaborates in the field with the elements and the features of drylands. The successive failures to establish *Faidherbia* plantations tend to blame irregular weather patterns, aggressive grazers, unsanctioned felling, insects, and other unpredictable dynamics. As blame is recast away from the human, it befalls the uncontrollable elements of the environment, often providing the grounds for scientific misinformation. Blame is certainly not reoriented to the plant, because it might be construed as a questioning of afforestation. To reorient responsibility would necessitate an acknowledgment that the plant is alive, behaving and reacting to the experiment that was so boldly and bravely funded. Is the living plant resisting conformity? Is it possible that taproots are sensing the presence of extreme competition evident in the density and scale of plantations?

An expression of collaboration between human and plant life influences diverse practices and produces inclusive methods. Outcomes might pressure sedentary agriculture into more fitting lands, further afield, and embrace nomadic traditions. The cycles of transformation embedded in the reverse deciduous display of *Faidherbia* and the deep-rooting tendency of its taproot are predictable behaviors that are selectively exploited or disregarded by early attempts to settle nomadic patterns and encourage pastoralism. These exchanges highlight how difficult it is to talk about plants in singular terms. Plant life is multifarious and includes the fine connections and associations that plants engender *with* the living environment, including humans. *Faidherbia* played a starring role in the fuelwood crisis through the 1970s, and more recently it plays the part of ideal dryland plant to "solve" the threats of extreme drought and desiccation. As a result, it permeates the lists and indices that chart increases in timber yield or decreases in spread of the desert. Yet, even with decades of sustained popularity that foster taxonomic debates, this plant generates its own glossary of operations; it refuses to cooperate with the construction of artifact.

INDEX

The Prairie States Forestry Project imagined by the procedures of artifact, index, and trace. Image by the author.

Index refers to the accumulation and distribution of planting units. In its unit state, the plant is condensed into an afforestation statistic. Unit calculations enable economic valuation and commercial trade, overpowering aliveness. As a result, *index* explores the procedures that help distribute units across material, physical environments, as evidence to support planting radically disconnected from the landscape. The transmission of units is sanctioned by species selection, the ultimate specification of *index*. Consequently, afforestation encroaches upon dry-lands through the lists of sanctioned species, specific catalogs, and manuals of standardization that reorganize biotic formation and any practical or traditional knowledge between human and plant life. The perversion between the conviction of human calculation on the one hand and the behavior of plant life on the other provides the context for considering *index*, as planting is achieved by mobilizing millions and billions of tree units.

This section explores the rise of afforestation in the United States through the persuasions of *index*. The authority of *index* advances the Prairie States Forestry Project, laying claim to tree planting as a national effort. At the time, forestry emerges as a profession that trades in resource extraction and yield metrics. The ensuing accumulation of units correlates with the accumulation of land, revealing that tree planting and tree felling could operate across the whole land surface of the United States.

> The subject matter of the profession of forestry is equally distinct from street tree-planting on the one side and landscape architecture on the other. It has to do with wooded regions, with the productiveness of forests, chiefly through conservative lumbering, and, in the treeless parts of the United States, with planting for economic reasons. Except for a comparatively small area of desert land in the West, the whole land surface of the United States is included in the possible field of work for the forester.[1]

Prior to professionalization, the subject matter of forestry was limited to wooded regions, or forests. The territory was broad, sweeping, and often coastal. It was riverine and arborized. Multifarious as it was, the challenge of dryness presented a threat to American expansion. In order to avoid failure, a set of procedures was required to suppress human kinship and distance people from the land. Clearing and planting trees helps procure marginal zones and overcome indigenous claims or land disputes in order to secure ownership. Afforestation confirms that the sequence unfolds regardless of social and biotic context.

The plant sciences do not participate in moving plant life to the field. Practical, dirty endeavors are the rough, failure-prone planting techniques of forestry, horticulture, and the nursery industry. Thus, the interlocking procedures of forestry fill the gap between horticulture (practical) and botany (scientific), enabling a translation from scientific truth, to specialized list, and finally to the specifications of engineered soil and containerized hybrids. It is this disparity that empowers *index* as the link between the fields of knowledge production and the professions of practice.

Professionalization is replete with examples whereby the mediocre and the common are placated by the certainty of superior knowledge, and the confusing part is that this procedure is often reinforced by scientific truth. The "temple of science" is precisely what makes modern science unique, conflating professionalism with progress. For instance, provisions made for woody plant units produce an illiteracy of dryland plants. The critical power of *index* lies in this agreement: it can mobilize plant life for exploitation at the same time as it renders its laws necessary and absolute.

At times, *index* is the literal list generated by experts. At other times, it is a kind of preproduction, a process of beginning to fix elements and generate content that enables governments to administer treeless biomes. This is particularly pronounced in relation to securing territory by tree planting, as trees are cast into prearranged, authorized lists sanctioned by remote authorities. Afforestation is ultimately a design project, an act of horticulture that should not be confused with a scientifically informed ecological endeavor. As we have seen, understanding the dynamic foundations of large-scale afforestation campaigns is crucial for our future, and for renewing a contextual relationship between species. At stake are how the maneuvers endure to reduce and misunderstand plant life, a diminution that slips easily into the norms of global aid, ecological rhetoric, and conservation mandates.

4

Confronting Treelessness

As I looked back at the results of my day's work, my spirits rose; for in the east a man might have worked all summer long to clear as much land as I had prepared for a crop on that first day. This morning it had been wilderness; now it was a field. . . . Surely this was a new world! Surely this was a world in which a man with the will to do might make something of himself. No waiting for the long processes by which the forests were reclaimed; but a new world with new processes, new neighbors, new ideas, new opportunities, new victories easily gained.

—Herbert Quick, *Vandermark's Folly*

IN PIONEERING THE MOVE WESTWARD, America expansionists emerged from the trials of their journey into a surprisingly treeless environment. While the protocols of clearing and cutting forest to make space for cultivation was a rehearsed practice of early European settlement on the East Coast, there was no precedent for the vast prairie landscape. By the nineteenth century, restrictive boundaries were already established for indigenous tribes in tandem with expanding ownership boundaries for Euro-Americans. The majority of this chapter is underwritten by a series of alienation acts that advance across the landscape prior to the national effort of afforestation under consideration. The sequential change in human occupancy of the prairie biome is reflected in the processes of planting—in what species are being planted, by whom, and where.

The region of the American prairie states is also referred to as the Great Plains, or the Great American Desert, and includes parts of Texas, Oklahoma, Kansas, Nebraska, South Dakota, and North Dakota. Ecologically, prairie ecosystems are a type of grassland biome. They are characterized by grasses (Poaceae) and forbs (nonwoody plants) of various heights and densities.[1] This "new world," as Herbert Quick's comment reiterates, was romanticized by settlers through its ease of cultivation. Since trees no longer had to be felled or cleared to create arable land, they would need to be planted in order to protect it.[2] Thus conceived, the presence or absence of trees lay in the structure for designing tree-planting protocols across the American prairie states regardless of the cultures that thrived for thousands of years. Likewise, the deep, ancient organisms that articulate as grasslands were invisible to speculators.

At the time, farming signified a potent form of nationalism, as American-owned territory expanded through the middle of the continent, advancing the spread of corn and wheat. The suggestion is that settler colonialism relies on tree planting. Planting came in two forms: crops as a duty of expansionism, and tree seeds from cherished specimens left behind in the eastern states. Either way, planting was an antidote to treelessness. If machinery was the new process, and agriculture was the new industry, then the triumph of settlement had three adjacent operations: build a home, raise a family, and plant trees. On the surface, it seemed that nothing could limit the procedures and protocols that linked planting with endless bounty and a pioneering spirit.[3] In this context, the potential of tall, vertical trees mitigated the endlessly horizontal fields and the wide plains around each cultivation effort. But the surface was not the issue. Treelessness represented freedom and assumed no rules, no fences, and no boundaries. As plots were quickly carved into centuries of prairie strata, the plow became the new metric of progress, as the novel *Vandermark's Folly* goes on to explain.

> As the team steadily pulled the machine along, I heard a curious thrilling sound as the knife went through the roots, a sort of murmuring as of protest at this violation.[4]

Farmers were equipped with machines and armed with botanical knowledge, as they actively territorialized endlessly horizontal fields of grasses and forbs for the sake of planting crops sanctioned by the newly formed Department of Agriculture.[5] This will be explained further. For now, imagine only that in pioneering the move westward, industrious farmers were actually confronting the concealed axis of prairie formation: the depth, weight, and volume of dryland roots and rhizomes.

The landscape of the prairie grasslands is commonly referred to as *vast,* an attribute of drylands exemplified by images of sweeping, boundless fields; gusts of erosive weather; and ranges of megafauna. The prairie itself is less often recognized, as it forms the backdrop of the predominant narrative of shifting weather patterns and the more recent hobbyist appreciation of wildlife. Underfoot, the fibrous roots of plants proceed in relation to a relatively sparse distribution of woody roots. All land plants have fibrous roots, from the agricultural crops we eat through global grasslands and the saplings of the forest. Of note is simply that trees have fibrous roots *and* woody roots.

Grasslands are composed of mainly fibrous roots, a composition that finds balance and collaboration where woody roots also emerge. Woody roots and rhizomes tend to grow along streambeds, tracing riverine fringes and clustering around surface waters. Grasslands, such as the ones that comprise the American prairies, have been likened to an organism, as grasses, forbs, and occasional woody stems develop along the surface of the earth and collaborate or compete through cycles of reproduction and dormancy.[6] Each belowground ensemble is coordinated with varying moisture regimes, which ensures aboveground presence, absence, and plasticity. Grasses are thickly intertwined, mingled, and associated with the climate. In the middle of the continent, the temperature shifts with reduced rainfall events. Tall woody plants do not adapt well to long periods of exposure and the climatic forces common to the region, especially water scarcity and high wind velocity. This is why they tend to hug water tables.

The coevolution of dryland plants and nomadic culture is particularly significant to a reading of the prairies as an organism. Richard Manning skillfully includes the grasses themselves when describing the iterant movement of organisms.

> Nomadism implies motion, not a random motion of the nomads' choosing, a wanderlust, but a motion dictated by the conditions of the land. Nomadic economy is based on wild grasses too thin to support pastoralism, the fixed pastures of domestic animals, or the commons that grew up in conjunction with settled agriculture. Pastoralism, the intensive raising of domestic stock, is an intermediate step between farming and nomadism. The nomadic economy is extensive rather than intensive. Nomads follow animals following grass.[7]

Consider the relationships embedded in the subtotals of each professional *index* and the irreducible temporality and rhythm of the biotic landscape. A relationship with time not only postpones and disrupts the scientific requisite to invent, assert, project, and distinguish but also expresses the ranges and cycles that are difficult to define as delineated territory.

Manning describes the "sweep" of drought as active expression of grassland formation "whereby the word balance seems wholly out of place."[8] The American grasslands undergo drought in thirty-year cycles, manifest as a fluctuating border between tallgrass prairie and shortgrass prairie. Grasses tend to retreat into their roots and the scattered trees thicken their bark, but all plants migrate to find less extreme ground, and more moisture. This motion in time and space is conscious survival of the prairie organism, where soil, plant, and climate collaborate from particle to particle, adhering, clustering, and forming relationships. Through cycles of aridity, drylands such as prairies are constantly rearranging, as succession confirms the process of renewal after drought, fire, or storms. The surface of the earth appears barren because the prairie actually rebuilds itself entirely through the relations in the rhizosphere.

Planting is the protocol of treelessness. Vast, naked, barren land was not deemed valuable by the farmer-developer equipped with an a priori theory of settlement. Instead, the prairies were cast as "a new world with new processes," embodying the sense of freedom experienced by speculators moving westward. The "new" processes advocated plowing and planting species that were scientifically proven to be valuable to human settlement. The plants of the prairie organism had no immediate value.

Consequently, each operation hinged on swapping out *unproductive* plants for productive ones. The inscription is almost too complicated to follow, as the plant is so deeply ingrained in our social and cultural consciousness that any activity and life displayed outside utility is unconscionable or seems irrelevant. Thus, triumph over the land was conceived through the success or failure of plants to adapt to human instruction and protocol. In order to encourage tree planting and address treelessness, the success of *index* is confirmed by newly established federal regulation, as rights to ownership were attained through records of tree planting.

Tree planting is not only a counterweight to treelessness; it stabilizes deforestation. What is rearranged through afforestation practices lays the foundation for the means and methods to plant over drylands in order to tally units. What is significant to these maneuvers is that while afforestation emerges at the continental scale through early American settlement, this history is not a remnant of turn-of-the-century resource extraction but endures as a form of environmentalism through coordinated campaigns that still drive global planting policy.

TIMBERING CULTURE

1862 Established. Department of Agriculture (USDA)

1873 Passed. Timber Culture Act

1875 Established. American Forestry Association (AFA)

1876 Established. Office of Forestry (USDA)

1880 Published. Report on the Forests of North America

1881 Renamed. Office of Forestry becomes Division of Forestry

1890 Passed. Nebraska Sandhills Afforestation Project

1891 Repealed. Timber Culture Act

The Timber Culture Act (1873) was a radical means to secure resources in otherwise treeless environments. Its stated claim provided entry "for the cultivation of timber which are prairie lands or other lands devoid of timber."[9] Its passing mobilized the protocols and promises of affidavit to deed land across the prairie states. The act was followed by land entries and land patents, the final steps in securing land that had already been peopled for thousands of years by hundreds of Native American tribes.[10] Nebraska—a prairie state—took the lead in land entry numbers, tallies, and acres, which frames the itinerary of tree planting as a form of land tenure predicated on afforestation because Nebraska is principally a dry prairie grassland.[11] Therefore, the Timber Culture Act sanctioned the substitution and deletion of dryland ecology, in this case both the extant prairie biome and the skillful cultivation of its inhabitants. A quick review of the associated institutional language makes evident the use of authority, hinged on administrative tasks. It further implicates the individual farmer as a developer, invested in the act of swapping thick grassy cover for individual tree units in order to establish an enterprise.

> Where 160 acres are taken, at least five must be plowed within one year from date of entry. The following, or second year, said five acres must be actually cultivated to crops or otherwise, and another five acres must be plowed. The third year the first five acres must be planted to trees, tree seeds or cuttings, and the second five acres actually cultivated to crop or otherwise. The fourth year, the second five acres must be planted to trees, tree seeds or cuttings, making, at the end of the fourth year, ten acres thus planted to trees.[12]

A "tree claim" could be filled in by anyone, and in return no more than 160 acres was acquired, with the specification that 40 acres of the claim must be tree planted, an effective means to ensure settlement. According to one claimant, "the section of land specified in my said application is composed exclusively of prairie lands, or other lands devoid of timber; that this filing and entry is made for the cultivation of timber and that I have made said application in good faith."[13] Units coalesced local and regional affairs, advancing federal and legislative ambitions. But the cost was only refined in the fine print, which specified that no fewer than 2,700 tree units could be planted on any ten-acre block.[14] It was a staggering number that led to copious tallies, monitoring programs, and the need for reportage. Each territorial procedure (survey, reportage) reinforced the hopelessness of the task: water was scarce, winds were strong, erosion seemed erratic, and the quantity of units could not be sustained.

A cluster of claims filed by a group or a household was one of the common characteristics of farm and ranch development. These clusters mark the distinction between fertile river valleys and the associated loess lands on the margins of the Sandhills environment. The physical landscape of the Sandhills was a challenge, and most claims failed.[15] To plow and maintain 160 acres with a mix of crops and trees was a challenge that took hold for very few claims. Managing 675 living trees on each acre was akin to owning a product, and growing trees was the best way to hold onto your land and avoid the cancelation of your claim. The number of entries made under the Timber Culture Act records area in acres, although territory could only be claimed if individual tree units were calculated and submitted for approval (see Table 2). This kind of accounting leaves out more than it includes, a discreet feature of *index,* marking both the beginning of afforestation-linked landscape management and the rise of incentive-based policy maladapted to drylands.

Table 2. Number of Entries Made under the Timber Culture Act

States and territories	1873		1874		1875		1876		1877		1878	
	No. of entries	Acres	No. of entries	Acres	No. of entries	Acres	No. of entries	Acres	No. of entries	Acres	No. of entries	Acres
Arizona	–	–	2	196.51	2	320.10	10	1,197.15	21	2,440.00	11	1,600.00
Arkansas	–	–	–	–	–	–	3	231.92	–	–	–	–
California	2	329.75	59	8,878.06	195	20,065.23	136	20,524.33	75	10,586.05	60	8,029.42
Colorado	–	–	17	2,272.24	27	3,454.82	45	6,514.22	28	3,343.33	125	17,436.73
Dakota	24	3,560.00	856	124,997.29	451	61,969.75	842	119,835.23	476	68,266.92	3,769	579,804.04
Idaho	–	–	2	180.83	21	2,583.25	17	1,973.89	52	7,035.91	158	22,169.53
Iowa	1	145.90	33	3,816.05	92	9,127.52	99	8,563.42	59	4,791.55	89	7,535.47
Kansas	60	9,642.17	1,954	282,479.07	1,265	168,269.06	1,354	185,596.43	1,666	238,020.44	4,031	593,295.17
Minnesota	95	14,710.15	804	113,131.63	499	63,673.73	1,070	140,126.30	561	76,021.53	2,693	377,017.78
Montana	–	–	–	–	–	–	–	–	3	398.59	9	960.00
Nebraska	137	21,858.07	2,164	312,712.09	1,061	130,894.26	834	106,499.74	706	90,812.90	1,408	195,306.68
Nevada	–	–	–	–	–	–	–	–	2	240.00	5	600.00
N. Mexico	–	–	–	–	–	–	7	1,128.00	–	–	2	320.00
Oregon	–	–	–	–	7	882.68	13	1,793.18	19	2,509.37	130	18,446.21
Utah	–	–	–	–	–	–	3	399.88	3	338.50	9	1,280.00
Washington	–	–	22	2,482.22	31	3,324.14	54	5,374.28	148	19,746.75	562	78,237.00
Wyoming	–	–	1	80.00	1	130.47	1	160.00	–	–	–	–
TOTAL	319	50,246.04	5,923	851,225.99	3,652	473,694.31	4,488	599,917.97	3,819	524,551.85	13,061	1,902,038.03

Source: Timber Culture Act, March 3, 1873, and June 14, 1878, to June 30, 1882, inclusive. Reproduced by permission of U.S. National Archives.

The Timber Culture Act was repealed less than twenty years after it was first legislated due to low survival rates. For instance, in Nebraska alone almost nine million acres were claimed under the act, but final proof was only made for trees planted across two million acres.[16] Since alterations to the land were less predictable than imagined, the gradual accumulation of expertise was necessary to further instill confidence in afforestation. The newly formed federal government certainly did not want western expansion to be deemed a failure.

Not surprisingly, this tension between success and failure required expertise, which arrived in the form of technical manuals, suitable species lists, and machinery types: the *index* of expertise. In the radical ecology of afforestation projects, the concept of failure is critical to the vocabulary of management, whereby catastrophe becomes a claim made by a specialist rather than a circumstance of ill-aligned biotic and climatic association. From a different angle, this means that expertise reinforces ecological projections through a series of calculated procedures rather than distinguishes, decodes, and evolves through direct experience with plant life. In fact, the plant really had nothing to do with it. Thus, failures in tallying units set the stage for the assimilation of expertise, which was equally an opportunity for the emerging business of arboriculture to become the science of forestry.

The Timber Culture Act mobilized substantial takings across drylands, revealing that trees were valued as an important natural resource before forestry emerged as a discipline. At the time, forestry was a colonial economy, a means to import and spread viable European plants across the American continent and send New World novelties back to Europe.[17] The botanical studies and the natural sciences were far too remote from the actual efforts of farming, explicated by the disparity created between increasing knowledge of plant parts in botany and the transmission of their application into common use, through horticulture. The more practical "cultures"—including the planting of trees in cities, crops for a homestead, or flowers for public display—were the domain of arboriculture, horticulture, or agriculture, for instance. Botanical studies were firmly indifferent to practical endeavors in much the same ways as farmers or ranchers are indifferent to scientific study. Botany's elevated status ensured that it did not labor, and its scientific status protected it from ever having to do so. So long as these two practices (botany and horticulture) remain entirely distinct and in organized separation, a summarized reading of plant life is multiplied into the space of physical, earthly matters.

The failure of the Timber Culture Act was not attributed to the ill calibration of species, quantity, acreage, rainfall, and temperature, or to the insistence of inserting trees into a vast, treeless landscape. Imagine a world of individual farmers working across vastly differentiated territories. This struggle to hold boundaries across a landscape without trees was a dynamic challenge. Consequently, the struggle to establish trees compelled professionalization. Failure was pinned on individual farmers and expertly manipulated as a means to assert more control over territory.[18] From the shady groves of the East Coast, the American drylands represented a challenge that could be overcome by new legislation, as increased federal regulation was imposed with publications of know-how from expert foresters.

Timber and crops were needed to support a growing population, and the prairies represented a blank canvas. As a result, tree culture across the Plains was predicated on the need for protection from the harsh climatic conditions, which offered a correspondingly reduced description of the grassland biome.[19] In order to increase agricultural production by sheltering farmlands from strong winds, each belt, break, or wall represented more than local shelter, firewood, or fuel: it was an opportunity to nationalize timber. The replicable model of planting a shelterbelt catalyzes the farmer as an agent of the federal economy by suppressing indigenous practices and local acquaintance. The language of this authority eliminates conjecture through levels of survey, reportage, and authoritative lists that impress with calculation.

> Of the total area (Great Plains region), some 56 percent lends itself to shelterbelt planting, about 39 percent is difficult to plant, and 5 percent is entirely unfit for planting.[20]

According to this federal explanation, there is nothing essential, intermediary, or material to constrain the accumulation of units. This description instrumentalized efforts to liberate the vacant spaces between meridians.

Settlement necessitated clearing and cutting forest to make space for cultivation, as the character of cheap nature ultimately supported expansion. Both tree planting and tree felling were exploitative regimes that consolidated control across the American continent. As we have seen, the difference was contextual; trees were cleared to make space for cultivation along the arborized East Coast, and trees were planted as pioneers moved west into the treeless prairies to cultivate land. In both cases, trees with clear human advantage were selected, while plants that grew of their own accord, or spontaneously, were overlooked and rarely stabilized in the procedures of *index*. It took decades of energy, planting, and felling before allegiance to the spontaneous plants was described by Charles Sprague Sargent.

> Many years ago, when I first realized the difficulty of obtaining any true knowledge of the trees in this country, I formed the plan of writing a Silva which should contain an account of all the species that grow spontaneously in the forests of North America.[21]

The study of trees as spontaneous organisms with indigenous characteristics was described in *The Silva of North America* (1891), a work of botanical inquiry that detailed 412 species, with reference to growth habit linked to illustrated maps that indicated distribution.[22] Most significantly, Sargent, the director of the newly established Arnold Arboretum at Harvard University, made some notes on the characteristics of wood. The breadth of Sargent's ambition is explained in the preface to the first edition, including his most pressing articulation of plant life, through engagement with "living states."

> To be really understood, they must be studied in the forest; and therefore, since the plan of writing this Silva was formed, I have examined the trees of America growing in their native homes from Canada to the banks of the Rio Grande and the Mountains of Arizona, and from British Columbia to the islands of Southern Florida. I have watched many of them in gardens of this country and in those of Europe and there are now hardly half a dozen of the trees which will be described in this work which I have not seen in a living state.[23]

At the time, Sargent held the position of professor of arboriculture, which ensured that his tenure at the arboretum would be focused on applied engagement with the landscape, with "plants in a living state." Sargent valued the associations between the individual tree, and its necessary relationship with soil, climate, and human codependency. His treatise provided a careful account of spontaneous woody plants, distinguishing between rainfall patterns and elevation, factors that limit tree growth at the 100th meridian. This is the boundary of aridity that reveals the prairies as dryland.

The variable edge is endowed with a watercolor-like transparency that emulates this bond between climate and plant life. Armed with the notes on "wood producing capacity," Sargent's treatise would go on to become the first technical forestry report of the newly created Division of Forestry in the Department of Agriculture in 1886. *The Silva of North America* elevates claims in drylands, as forestry reaches into scientific study to nurture its advance. Of note is that while the profession of forestry is confirmed in the United States in the European tradition, it expands to include planting in order to secure extractable resources. The varied institutional settings between botany as a science and forestry as a profession entangle the study of plant life in the discrepancies and difficulties reports, as forestry read gaps in the maps and botanists read their extents. Rather than carefully consider the findings of Sargent's botanical fieldwork, remote structures deployed the literal blank space of his mapping efforts as a means to recontextualize the prairies as a tree-planting endeavor, converting biomes and replacing thick, fibrous rhizomes with woody plant units.

Index is a method of standardization that fine-tunes data to suit industrial expertise. As a case in point, the appropriations of forestry transformed *Silva* into an operations manual—or a masterplan—which begins to explain how vast quantities of grassland expanse and ancient reserves of pastoral acreage become the jurisdiction of federal foresters. The map was "empty," a curse of drylands the world over, but it was an opportunity for the industrious. The insertion of trees infused meaning into New World expansion in the prairie states, as demand for ownership lines and boundaries enabled land claims hinged on tree planting. Afforestation emerges as forestry (not botany) and disseminates plant life as a "tree unit," transferring knowledge of resource management to each small-hold farmer. In this way, particularly profitable and predictable plant species are endorsed over others, superimposing units on extant plant life.

The adoption of horticultural techniques such as plant selection helped convert, confirm, and manage vast ecological dynamics. Detailed practices involve production and procurement of seed and the circulation of seedlings, moving plant life from scientific theory into spatial practice as presence or absence of exploitable units enriches territorial claims. Selection also fundamentally preferences certain plant-based resources over the living prairie organism.

Afforestation relies on plant-as-artifact in order to calculate and tally through *index,* attesting to the reliance on but ultimate exploitation of botany. In the succeeding section, tree experts and forestry professionals are surveyed in the context of the American grassland prairies, illustrating the inventiveness of *index,* whereby citizens and communities contribute to a national policy of industrializing biota by supporting continental tree-planting programs. In advance, the study of roots and their relations in the prairie grasslands provides contrast to the cheap nature of the plant unit. A consideration of the relations that resist the procedures of *index* creates a distinct ontological zone: of plant life on the one hand and human calculation on the other.

EXCAVATING PLANT ECOLOGY

Plant ecology originates among a lineage of scientists and scholars like Sargent who contemplated landscape dynamics outside exploitative practices and the remote authority that underscores resource extraction. At the turn of the century, American botanists embraced the elaboration of ecology developing in Europe, increasing the study of individual plants prior to the establishment of the office of forestry.[24] While the acceptance of larger biotic connectivity was practiced by plant botanists, their work found designation. "Plant ecology" is defined as the living plant studied through origin, development, and structure, through both time and process.[25] Notable plant ecologists emerged from federally sponsored land-grant universities, stimulated by a close association with their climatic location: Emmy Lucy Braun in the eastern deciduous forest; Forrest Shreve in the deserts of the Southwest; Henry Chandler Cowles in the Lake Michigan dunes, and the scholarship of the grassland scientists Frederic E. Clements,

John E. Weaver, and Homer LeRoy Shantz in the prairie states.[26] At issue was not where the plant was located, or how it could be exploited, but what activities and progression were taking place, including how plant life moves and responds to environmental change. Plants were the site of speculation in the production of knowledge that coalesced philosophical estimation and practical fieldwork. The origins of American ecology progressed with attention to plants. Specifically, plant ecology advanced in the study of prairie plants.

Prairie is not a type of vegetation; it is an organic entity established in collaboration with the soil and the climate over centuries. Prairies form biotically in grasslands, along with savannas, scrublands, pampas, and steppes. In ecological terms, grasslands are referred to as a formation, implying that the landscape is in a constant state of materialization, reliant on periods of intense dormancy; for half a year, the living prairie is underground.[27] Thus, both decay and persistence are essential features of prairie formation. Its surface enlarges and shrinks, depending on signals from continuous manufacture in the subsurface. Extensive grasslands occur on every continent, marked only by strongly differentiated wet and dry seasons where rainfall is too low to support a forest but higher than that which results in desert life-forms. As a result, grassland formations can be found in subtropical to temperate ranges, from the African sub-Sahel to coastal Venezuela. The obvious, outstanding feature of any grassland is its extensive horizontality, as it presses its bulk on the surface of the territory, resulting in few gaps where tree cover may emerge. Typically, scrubby tree silhouettes and vast open horizons characterize views. As a result, no other biome has been as easy to cultivate, destroy, or replace.

Plant life permeates and saturates each fine horizon of soil within the rhizosphere, outpacing the spread of visible or aboveground portions. The most remarkable interactions occur in this rhizography.[28] Woody and nonwoody root systems demonstrate varying relationships, associations, and tendencies, but it is the behavior of fibrous grasses that presents *depth* as a specialized adaptation. Investigations of the grasses and sedges in the American prairies confirm that profound depths are the rule among grassland species. Frequent grazing, burning, or damage may result in the loss of the singular leaf blades, but this disturbance produces more penetration in the rhizosphere since the leaf growth zone (transverse intercalary) is located in the lower portion of the shoot, closest to the primary root. Disturbance also adds density to the roots, as they weave through a web of *mycorrhizae*—a word derived from the Greek *Mykyes,* meaning "fungus," and *Rhiza,* "root"—a microorganism that links the plant to the earth.

The roots and rhizomes of grasses and forbs construct the prairie landscape, confirming that plant *living* is done underground. Thus the foundations of American ecology are linked not only to the plant but to plant roots and rhizomes. According to the work of American grassland scientist John E. Weaver (1884–1966) at the University of Nebraska, there are particular layers upon which species prosper in relation to varying volumes of organic matter, microorganisms, and nutrients in the substrate.[29] Some roots extend only two feet deep and grow denser, while others reach four feet and then grow in density. A fine balance of absorptive capacity and miles of root length determines the physical conditions in this threaded universe.

> So many species—often a total of 200 or more per square mile—can exist together only by sharing the soil at different levels, by obtaining light at different heights, and by making maximum demands for water, nutrients, and light at different seasons of the year. Legumes add nitrogen to the soil; tall plants protect the lower ones from the heating and drying effects of full isolation; and the mat-formers and other prostrate species further reduce water loss by covering the soil's surface, living in an atmosphere that is much better supplied with moisture than are the windswept plants above them.[30]

As different root systems aspire to absorb nutrients and water, a sophisticated response in depth relations ensues. The prairie formation produces plants with infinite life spans as non-woody roots spread meters underground, reproducing, competing, and surviving in an obscured vertical hierarchy. Weaver's experiments reveal horizons of more than thirty feet in depth.[31] This vertical layering is effectively how grassland plants vie for survival but do not outcompete each other. Weaver's suggestion of the "segregation" of the soil is actually part of the ongoing modification between prairie plants and their above- and belowground environments, which operate more territorially below than above the surface. Unlike woody roots—which penetrate the soil and seek nutrients—fibrous roots fully occupy their opaque environment and absorb nutrients in collaboration.

Measuring depth and density in the geography of dryland soil presents undesirable conditions for the human scientist, necessitating comparison. In his efforts to interpret the habit of roots, Weaver compared underground systems to what he called "familiar aboveground parts."[32] Still, he sought to make his discoveries palatable by transforming his approach; Weaver did not build a laboratory ex situ but excavated a laboratory in situ. Weaver's grassland laboratory was unearthed as an excavated four-sided room that he reached by descending a ladder. Although it was a laborious technique, facing the rhizosphere freed his experiments and allowed his plants to be to excavated through the walls, gently revealing interlocking roots with a hand pick and a dry paint brush. To ensure certainty, Weaver also found that when the root system was excavated and photographed, many of the finer branches and root ends were concealed or desiccated. In seeking greater precision, Weaver turned to drawing. He rendered his root excavations to scale on a large sheet of grid paper, taking advantage of firsthand interpretation. Descriptions of the prairie formation progressed through direct engagement with the field using the living rhizography to collectively augment knowledge of plant life in ecological terms.

In order to examine root production and depth relations, curious and resourceful scientists like Weaver put themselves in their experiments, penetrating the field itself in order to observe living root systems. For the first time, the rhizosphere was not a hopeless dilemma to be cast out of sight or replaced by desiccated parts. The prairie formation emerged as an unknown, underground universe worthy of further scholarship. Studies achieved comparison and direct experience, not only in pure stands but also across vast formations, from Oklahoma's true prairies to Nebraska's Sandhills. Identifying strata necessitated living plants and extensive fieldwork, methods that underscored Weaver's experimental approach. In so doing, static artifacts were replaced by his considerations of the growing plant. Weaver's botanic research embraces the behavior of the individual plant organism, while his plant ecology defined geographic territory from the ground up.

Weaver advanced the study of plant ecology through novel engagement that emerged in the relationship between the plant and the climate, the soil and the organism. His work flourished within the collaborative environment of the Department of Botany at the University of Nebraska.[33] His advisor and contemporary was Frederic Clements (1874–1945), an early proponent of plant ecology who would later join the newly formed U.S. Soil Conservation Service. Clements advanced his botanic scholarship by emphasizing vegetation—as opposed to plants— at the scale of environmental systems. By determining the presence of certain plants as indicators of larger areas (and thus the absence of others), Clements deftly transformed botanical interest in the individual plant into a concern for vegetation as a unit. Even in his early work at the turn of the century, nonstatic concepts at the scale of the landscape were inspired by botanical speculation.

> It is entirely superfluous to speak of dynamic and static effects of the plant, and the use of these terms with reference to the formation becomes equally unnecessary as soon as the latter is looked upon as an organism.[34]

Weaver and Clements liberated plant life by designating the entire prairie landscape as an organism, an entity that exchanges, perishes, and prospers in collaboration with the climate.[35] Plant ecology is what Sargent refers to as a living state. Such a thesis resonates with how plants adapt to earthly changes. For instance, the plants of the Great Plains went from tundra immediately after the withdrawal of the ice sheets five thousand years ago to prairie in warm dry periods and spruce forest in wetter times.[36] The methods that Weaver developed mobilized layers of biotic history in order to explain the significance of drylands.

Plant ecology reminds us that not all scientific endeavors target economic outcomes and that developing methods to include the entire living plant integrates the human intellect into the system under consideration. Plant aliveness includes transformation and spontaneity, as plant life is enthusiastically embraced without direct political or capital outcomes. Here, the human scientist works *with* the agency of the living environment. The differences are embedded in the aim of going out into the world to describe it rather than going out into the world to control it. One operates with a vocabulary of authority and the other, curiosity.

The methods of early plant ecologists create friction between the demands of applied and pure approaches to scientific studies, the "hard" and "soft" sciences. Although Weaver and his colleagues were applying scientific calculation, they were experimenting with norms. While the origins of plant ecology are embedded in the deep relationship between prairie grasses and forbs, the outcomes were not products, and they did little to support expansionist agendas. Absent any practical applications, grassland science had no direct benefit for the rapid urbanization of the soil or the development of forestry practices. The grassland scientists were meddling in the soil, in a world where the aboveground plant is valued as a resource. As systems came to replace individuals though ecology, the plant as grass, shrub, or tree is almost entirely erased, or replaced by the term *vegetation*. Absent behavior, plant life is distorted into ecology away from a biotic process, into a box on a cartographic legend, which was an ideal standard for foresters. Subsequent distortions of land typically indicated forest versus nonforest cover, an itinerary that resonated with the visible deforestation and depletion unfolding across the arborized East Coast.

NEW WORLD FOREST PROBLEMS

1891 Published. *The Silva of North America* (Sargent)
1901 Renamed. Division of Forestry becomes the Bureau of Forestry
1905 Renamed. Bureau of Forestry becomes Forest Service
1910 Published. *The Fight for Conservation* (Pinchot)

In the American consciousness, the origins of environmentalism are tied to the ethics of conservation, first recognized as *forest* conservation. The lineage of conservation, like many historic accounts, is entirely dependent on the narrator. For instance, ecological thought in America was first affirmed as *plant* ecology, the study of plant life in the environment. What is worth bearing in mind is that despite its significance, plant ecology did not proceed as the science of consequence in the grassland biome. Rather, forestry administered and studied the prairie formation. The reasons for this odd fragmentation of knowledge are slippery to pin down, but it is worth noting the role of Gifford Pinchot (1865–1946). Pinchot was more of a politician than a forester, and he acted as the first head of the newly established Forest

Service in 1905. Under his leadership, forestry progressed from a division within the Department of Agriculture (USDA) to a separate bureau responsible for forest policy across the whole land surface of the United States. The Forest Service emerged—not coincidentally—as the nation demanded unprecedented quantities of timber for fuel and construction. Charged with a simplified problem statement, Pinchot pushed forest conservation into unchartered territory.

The influence of George Perkins Marsh (1801–82) motivated Pinchot's interest in the debates of conservation. In *Man and Nature* (1864), Marsh determined that humans were a major geologic influence, famously asserting that "man is everywhere a disturbing agent," a landmark statement in both ecological and conservation studies.[37] Marsh was neither forester nor plant ecologist but articulated an unparalleled understanding of the role of plant life in preserving the landscape, in how a plant's connection to the soil maintains the substrate and prevents or reduces erosion. According to biographer David Lowenthal, the success of Marsh's rhetoric is found in his caution concerning impending disaster, by addressing what he termed "New World" forest problems. Marsh challenged the general belief that human impact on nature was generally negligible, explicating a global tendency toward wasted natural resources. Armed with *Man and Nature*, Pinchot adopts conservation as forest policy. In a sticky exchange of authority, conservation and forestry coalesce as both ethic and science, alternatively sanctioning felling and planting trees regardless of context.

Despite Marsh's warning, the Forest Service appropriated conservation as a "solution" to forest devastation, seizing upon Marsh's warning of old-world forest decline as a motivation to enforce national policy. Conservation policy progressed through extant forests and forest plantations, as plant ecology evolved as a scientific discipline in the prairie states. Remarkably, the difference was not correlated, although ecology revealed the role of plants in forming the territory and conservation uncovered the role of human agency over the territory. Rather, conservation was annexed to the profession of forestry, authorizing the Forest Service to oversee the planting of grasslands, prairies, agricultural lots, deserts, urban centers, and high mountains, concealing their extractive procedures in the outlines of conservation.

Tree-planting projects begin to materialize in the millions of acres as the soil is planted, replanted, aplanted, seeded, fenced, and labeled. The landscape is objectified, appearing and disappearing through measurement and calculation. In this way, felling and planting develop synonymously, rendering the environmental messages of *Man and Nature* were translated as generalized statements with gross ecological miscalculations. For instance, the notion that tree planting would bring rain to the arid West was appropriated to encourage settlers to move to ever-dryer lands.[38] The assumed property that followed tree planting is a clever disguise for afforestation. As Lowenthal's biography verifies, Marsh's message was pillaged for such "proof" that rainfall and prosperity followed tree planting. This nexus between national forest policy and conservationist thought is repeatedly stressed in Pinchot's *The Fight for Conservation* (1910).

> As a forester I am glad to believe that conservation began with forestry, and that the principles which govern the Forest Service in particular and forestry in general are also the ideas that control conservation.[39]

The fight for conservation was taken up by the Forest Service, exploiting the growing concern over *forest devastation* in order to plant and thus control more federal land.

Tree cover emerges in the lexicon of conservation. Land, absent "cover," is manipulated into land that can potentially receive cover. Tree units add up, when acres of "cover" are procedurally converted by the practices of forestry, as tree planting is sanctioned in non-treed environments. Consider the following passage.

> The crown has more to do with the life of the tree than its other parts, for the most important processes in the reproduction of the tree and the digestion of its food take place in the crown. For this reason, and because we can control its shape and size more easily and directly than that of the roots or trunk, the crown is of special interest to the forester. It is almost exclusively with the crowns that he has to deal in tending a crop of trees and preparing the way for succeeding generations. As they stand together in the forest, the crowns of trees form a broken shelter, which is usually spoken of as the leaf canopy, but which may be better called the *cover*.[40]

Land, absent "cover," represents millions of acres of potential federal territory. The conversion of "crown" or "leaf canopy" into "cover" had a number of spatial consequences that will be addressed further. For now, the consequences are especially poignant because the root system is set aside in the first pages. Regulators, policy makers, and the forest service administrators intentionally reduced the growing plant to standing timber. The consequence is that the federal government had no use for the nascent scholarship of grassland science or the complications of plant ecology. The hidden intelligence of rhizomatic formation could not reaffirm exploitation or capital return, one of the most fundamental reasons why canopy continues to prevail in carbon-capture calculations. This is what professional expertise does when it lays claim to fact and asserts the power of disqualification instead.

Pinchot's use of the term *cover* not only left out the aliveness of plants; it delineated the absence or presence of trees. The social implications and the negligence toward practical knowledge or public intelligence were already a form of expert persuasion upon which the profession of forestry is established: treeless lands can be conserved by planting trees. This raises another question: What are the spatial—and thus social—consequences of such abridged problem statements?

If Marsh explicated the "problems" of human action, Pinchot clarified that forestry had solutions, a defense strategy that he claims as a "virile evolution of the campaign for conservation of the nation's resources."[41] Conveniently, the solution relies on the interrelated goals of national expansion and resource extraction. By associating tree planting with conservation, Pinchot appealed to President Teddy Roosevelt, who pioneered land conservation armed with Pinchot's treatises.[42] Thus, a fragmented translation of *Man and Nature* is interpreted by statistical proof in Pinchot's *The Fight for Conservation*.

At the time of *Man and Nature*'s publication, federal regulations had already secured the first Forestry Service in the federal Department of Agriculture, established a forestry program at Yale University, and set aside more than 150 national forests as reserve areas. Such policy and regulatory potency have lasting spatial repercussions. Particular to the success of forestry was the reduction of national geography to manageable units, as conservation became a capital project.[43] Thus, the terms of conservation mitigated the affiliation between planting and felling, but conservation was fortified by yield management, the precision of *index*. This is significant to the ways in which management and development continue to prosper through the invention of crisis.

Specification in Units

Afforestation hides in the manipulation of specifications, loudly advocating for tree planting while quietly converting biomes, while there are no objective ways to measure the transformation, since partitioning the landscape relies on subtractive and additive procedures that operate in rotation. Cut here, fill there or fell here, plant there. The cyclical arrangement is wedged between industrial and ecological intentions, necessitating expertise in both analyzing extant lands and constructing entirely new ones. Not surprisingly, the interdependencies between forest and field mobilize the need for more and more units as a primary measure of converting biomes, or planting trees to secure timber. The forester does not deal in individual trees destined for a purpose but on masses of tree units, because the individual tree has value only as part of the whole.

At the same time that Weaver was excavating earthy workshops and painstakingly detailing the depth and intricacy of the prairie formation, the character of American expansionism was emerging in tree units, as domesticated plants balanced settlement. The accumulation and distribution of homogenous units triumphed over the diverse, interconnected, spontaneous, and multiple, preventing any expression from unscientific worlds and erasing the intimate aspects of daily life where social, cultural, environmental, and plant life mingle. Armed with an antidote to treelessness, forest policy worked its way across 170 million acres of American dryland by stabilizing a sanctioned list of tree varieties, despite the warnings that cycles that transform forest to field to plow flatten diversity and destroy soil by field wash at both an unprecedented scale and an unprecedented speed due to plow and policy. The Timber Culture Act confirmed that only the most reliable, predictable plants could be confidently inserted, so these updated lists confirmed only trees that could persist in prairie soil. To facilitate planting, the specification of tree units was linear, and in the American context catalyzed the shelterbelt typology.

Dryland shelterbelts are additive, unlike the shelterbelts in Europe or those found in the American East, which are primarily composed of remnant forest. Such belts are the residues of clearing, burning, or grading procedures. The plants that remain intact are often called hedgerows or windbreaks, and are registered vertically through a prolific, enduring accumulation of biota that remain deep in the rhizosphere. In other words, the trees, shrubs, and herbs of the former landscape are unbroken, and continue to thrive in diversity and habitat. As a network of accretive forces and biological activity, they are extremely effective at slowing erosive forces. The subtractive force of clearing land produces an increase in mounds and layers within the hedgerows, as branches and rocks are thrown into piles and endure. Each mature specimen also maintains vigorous seed banks and robust sprouting stumps.

When a plant is inserted as an individual organism, copied and pasted in neat rows, it has limited ability to produce relations in the soil. This is due to the constitution of the rhizosphere, in which a process between microorganisms, fungus, and soil biota is stratified in cooperation with other neighboring organisms. Plants, left behind with their major root systems and mycorrhizal relationships intact, continue to provide reliable growth patterns. The connection of each remnant to the soil advances the entire network, an interdependent and persistent spatial structure, which pulses through the soil. In contrast, shelterbelt protocols in drylands insert plant units into flattened, cleared, turned, and exposed soil. Plants—treated as objects—are forced to instigate each belowground relationship anew before any achievement can be registered aboveground.

A shelterbelt not only has behaviors, domains, and unknowns; it has a spatial geometry that often necessitates two landowners, as it straddles property lines. Further, this geometric configuration not only extends a thickened line and extrudes a height gradient; it extends the root zone in both the horizontal and vertical dimensions. The roots require irrigation, nutrition, and time in order to assert aboveground effects. As a result, local disputes escalate along with the behavior of any trees that display significant mobility, as some plants spread more quickly than others.

While a shelterbelt is designed to grow in height and girth, it is certainly not expected to migrate or move. If a shelterbelt is "successful," it is only because the plant evolved persistent root morphology and reproductive capacity, a feature that cannot be rendered visible or valuated economically. Yet conflict arises if the plant is successful enough to be found encroaching beyond property lines, or into productive farmland.[44] For instance, the behavior of some woody roots will outcompete annual crops for resources such as moisture and nutrients, which makes them a nuisance to farmers. In shelterbelt culture, trees are deemed beneficial only if they provide undeviating protection, and are deemed a menace if they overcome specifications.

The correlation between additive and subtracted spatial geometry illuminates the friction between developers and their environment during this period. Tree planting in the Plains exemplified this struggle, as failure to adapt was decidedly unacceptable and un-American.[45] Resolve only increases in the space between dying and trying, a feature of most developers bent on return. But it was the lack of consensus between investments that finally led to the appeal for government support.[46] Just as the American pioneers resisted treelessness, they could not cooperate on investment between property limits. Ownership is a personal discipline, as the contested fence and neighbor relationship suitably explicates. In a large-scale planted system, these thresholds prove mutually beneficial. Yet the owner who initially broke ground, transplanted, or seeded the rows tended to be the one to water, manage, and maintain it, despite shared outcomes. Effort was the developer's investment, and resistance has the power to affect the system it endorses.

Simultaneous to the decades when professional forestry, sedentary agriculture, and public acquisition were determining the American landscape, conservation collapsed neatly into agencies, services, institutes, and organizations. Political agendas merged with development strategies through the protection or abuse of natural resources, so that access and recreation became the most salient features of the vistas, lakes, peaks, and other wonders of natural beauty.[47] Forestry lurks in the background as the landscape becomes a cultural and social construct, devoid of individual life-forms beyond the human.

The symmetrical competition between growth and yield, or materializing and subtracting, currently calculates carbon credits through additive operations and timber volume in deductive increments. At once, local populations are led to believe that "billion tree" campaigns and "great green walls" are pronouncing progress, but growth statistics and depletion metrics are isolated from accountability and context. These flip-side tactics benefit from quick rotations that leverage tree planting as a means to elevate control through yield, in a world troubled by ongoing crisis.

When dealing in units, the authority of forestation selectively neglects the less visible attributes of plant life, including individual plant behavior, symbiotic relationships, and the concealed roots and rhizomes that form and deform the soil within which biomes are produced. Because of aboveground visibility, development budgets, and technical encouragement, projects that eagerly plant trees under the rubric of afforestation continue to garner international funds and approval despite the haunting conspiracy of replacing biomes and commercializing plant life. The prowess of progress that celebrates biome conversion creates as it destroys, a central tenet of afforestation.

5

Prairie States Forestry Project

THE PRAIRIE STATES FORESTRY PROJECT (1934–42) epitomizes the rise of continentally scaled afforestation campaigns. The project presents a centralized attempt by the U.S. Forest Service to unite the states, following significant social devastation marked by the Dust Bowl drought. The project, as an ecology of insertion, encompasses Texas, Oklahoma, Kansas, Nebraska, South Dakota, and North Dakota, also called the prairie states. In some ways the project displays impressive ingenuity following economic collapse, a kind of creative resourcefulness marked by President Franklin D. Roosevelt's New Deal policies. But the actual socio-ecological complications of the Prairie States Forestry Project suggest an unexpected tyranny that dominates dryness and erases any remnant of Native American titles by insisting on fixed frontiers. Rather than worry about the brutality of superimposing settlement, the project advanced by appealing to the malaise of losing one's home to "natural" disasters.

> Tree planting as a measure for improving Plains conditions has received attention since the advent of the early settlers. Trees grouped together as windbreaks or shelterbelts are credited with the improvement of physical conditions, probably most tangibly expressed as the protection of crops and cropland. A larger and more vital value, however, and one that cannot be expressed in physical terms or realized by those who have not experienced life in the prairie-plains region, is the reinforcement of the people's morale that comes with shade from sun glare, shelter from the ever-prevailing winds, the improved appearance of the countryside, a greater pride in ownership, and a real increase in value of the farmstead—all culminating in a general sense of being at home on the land.[1]

On the one hand, the Prairie States Forestry Project was an effective program of the New Deal, and well documented as a successful social agenda—creating jobs and reinvigorating the nation with optimism. On the other hand, it was a "natural" response to a "natural" disaster. As

geographer Neil Smith so aptly confirms, the "naturalness" of disasters is ideological camou-
flage for the *ruts* of social difference they encounter.[2] The Dust Bowl was not simply a disaster;
it was a manmade disaster, a disaster created by decree.

This is not only an ecological point; it is a practical one. The response to disaster typically
instigates a portfolio of interdependent policies. In this case, the historical particularity of the
program expands upon the protocols of introduction and assimilation developed during the
violence of settler colonialism, and the Indian Act. What binds species together is that the failures and conflicts of settlement are assuaged by plant introductions. The Prairie States Forestry
Project is fraught with questions concerning the "naturalness" of the investment, but in relation
to settler tactics it reiterates how citizens are armed with plant species as a means to assimilate,
to take hold of lands. Some species fight back; some die trying.

The persuasion of the project organizes around the growing evidence that making a permanent *home* in the prairies was actually more arduous than Congress imagined. Tree units were
drafted into service in order to prevent further migration *away* from state-sponsored settlement. Every unit specification engenders a conflict between trees and grasses, between nomadic
culture and developer temperaments, between plant life and human life.

Shelterbelt planting is tightly bound up with human struggle, as *improvement* is assumed
by the immediate need to protect homesteads from erosion. Similar reference to well-being is
claimed for resources including forest strips, farm grids, and hedgerows, trees grouped together.
Tree planting is justified by the difficult physical conditions, what would now be accepted as
the dynamics of the environment. Still, environmental conditions are never isolated, so improvement is always hinged on shifting relationships—human and otherwise. This is especially
the case when the environment resists improvement. The friction originates in the difference
between typology and ecology: between normalizing isolated units based on a predetermined
plan, and engendering relationships based on firsthand experience. Typology will resurface
as a professional technique in each case, a generalization of type that emerges regardless of
the resource rush under consideration. But typology finds particular resonance in the Prairie
States Forestry Project due to the rise of professional forestry and its questionable relationship
with unrolling typology as a technique of conservation.

The quotation near the beginning of this chapter is the first paragraph of *Possibilities of
Shelterbelt Planting in the Plains Region* (1935), a master-plan document compiled by the U.S.
Forest Service. The master plan is the ultimate *index,* consolidated by those in power. The document outlines funding provisions, location maps, climate charts, site requirements, and species
selection for each state along the ultimately unified extends of the project. Typology emerges
in the replacement strategy of plant introductions, as rows of tree units are depicted absent

context but within an abstract network. It is the ultimate planning document to shed light on the uncomfortable encounters often created by design.

The document triggered the popular imagination by depicting a verdant strip of trees within the otherwise treeless landscape. The message is that farmers must advance tree plant-ing across semiarid drylands in order to combat *unproductive* soil. The use of typology is paired with lists that describe particular tree units. Each row is specified perpendicular to the wind, across an area 100 miles wide and about 1,150 miles long, between the 98th and 100th meridi-ans.[3] The spectacular scale is associated with ingenuity; successful tree units create a noticeable localized effect. If trees develop suitably in drylands, they help alter wind patterns, protecting crops from hot winds in summer and breaking the cold blasts from the north in the winter. But, more often than not, tree units fail. If the plants did not fail on their own due to lack of water or stewardship, they tended to be destroyed by the extremes of the dry climate. Prairie fires rage across short grass country from lack of summer moisture; electric storms frighten spring's arrival with lurid flashes and thunder claps that make the earth tremble; severe win-ters are often followed by severe spring floods; and tornadoes sweep across the region, often forming in the Texas panhandle, traveling across Oklahoma and central Kansas, and frequently striking in Nebraska.

The Prairie States Forestry Project is so vast and each territory so alive that any attempt to describe the project risks simplification. Zooming in on the state of Nebraska suggests a means to sample the parameters, confrontations, and anomalies encountered across each state. Nebraska is a remarkable region, one in which the clock seems to be running backward as the population dwindles across rangelands. It is also home to a ninety-thousand-acre afforesta-tion effort, an entirely hand-planted national forest.[4] It is also the site of the first federal tree nursery, which explains why Nebraskan experience reiterates throughout *Possibilities of Shel-terbelt Planting in the Plains Region*. The nursery was both supply and evidence for the federal government, proof that afforestation could reboot the failure of settled agriculture across the prairie states.

The assumption that the failure of settlement can be offset by a series of tree lines or shel-terbelts minimizes the actual collision between enterprising activities and extant ecological niches. A shelterbelt might describe a "natural" approach to ease human discomfort and suffer-ing wrapped up in morale and pride, culminating in a general sense of being at home on the land, or it can be framed as a state-building project that marginalizes culture as it pacifies land, a common device of territorial dominance. Under the traditions of assimilation justified by early settlement, the federal government used tree planting as a distraction from the actual atroc-ities that were unfolding between humans. Tree planting furthered fixed settlement, tracing the linear margins between Anglo-American farmsteads and Indian territory.[5] This includes the friction between indigenous rights and expansionist policy as well as the struggle arising for newly settled farmers and their families.

The Prairie States Forestry Project embodies the politics of plant life because legislation remediated agricultural policy, not deteriorated prairie grasslands. Considering that the study of grassland ecology did not exist in China until the 1990s, and that most of Africa's Sahelian grasslands supported nomadic customs until mid-century French occupation, the American shelterbelt project represents the largest sanctioned attempt to uniformly transform drylands at a continental scale.

THE FIRST ARTIFICIAL NATIONAL FOREST

Nebraska is a state of forest firsts: the first state to propose Arbor Day (1872), the first state to champion the Timber Culture Act (1873), the first state to establish a federal tree nursery (1903), and the first state to plant an artificial national forest. Each episode in the history of state formation popularized tree cultivation under the rubric of conservation, without addressing the perilous forms of regulation that originated through forest policy. Issuing land rights to settlers materialized policy by subduing complex relationships between species. This arrangement is not only a historic account—a lesson of consequence—but echoes in each novel eco-designation that continues to delineate legal ownership in the twenty-first century. Even a quick search reveals that the stakes are federal, multinational, and global across the Nebraska prairies, ranging in support from the World Wildlife Fund (WWF) to the Environmental Protection Agency (EPA). Afforestation across Nebraska was taken to an extreme with the creation of an artificial forest. Nebraska provides a glimpse into the underpinnings of the Prairie States Forestry Project, a context that prepares the ground for the advance of tree units.

Nebraska also faced the first resistance to relocation assertions leading up to the Kansas-Nebraska Act (1854). Often called the "Nebraska Bill," the act effectively substituted one policy for another, upgrading politics to suit the resistance. In the name of law, two more states (Kansas and Nebraska) were impressed into the map, pushing fixed settlement deeper into the continent, while recomposing seasonal or iterant cultures. When policy rearranges territory, it guarantees that there are winners and losers. In this case, the Kansas-Nebraska Act invented the demand to organize the territory, in order to actually reorganize and consolidate "Indian" holdings, forcing tribes to cede their lands.[6] All remaining "Indian" lands would be ceded and the territory surveyed. The subsequent dispossession between Native Americans and Euro-Americans created enclosed reservations. Prior to the Kansas-Nebraska Act, state lines were not drawn, as features of the landscape delineated human movement, notably because a vast sand dune covers slightly more than one-fourth of present-day Nebraska.[7] These are the Sandhills, a region larger than Massachusetts, Connecticut, and Rhode Island combined. The dynamism of the Sandhills is shaped by the extensive ranges of annual grasses and seasonal hunting that does not align with the traditional conception of demarcated plantations or corralled livestock whereby resources are tallied. The Kansas-Nebraska Act is significant because it resolved the question of ownership: the new regional boundaries positioned the Sandhills entirely within the state of Nebraska.

The fine-tuned relationships between human and plant life were destroyed, forever lost as land was stripped away from Native Americans in central Nebraska. For instance, the Pawnee thrived across the landscape by listening to spring lightning as the first sign to plant crops, moving on through to follow bison herds, and leaving their crops to ripen unattended in the heat of summer to pursue the bison hunt.[8] In Nebraska, land claims sought permanent occupancy of the Sandhills by denying iterant dryland relationships.

> The Indians knew as well as anyone that if peace was accepted it would mean extinction, it would mean peace at a terrible cost, it would mean death and destruction and the end of the race. Their land was coveted and would sooner or later be taken.... Since the Sioux came into the possession of the Western country, which was ceded to them by the treaty of 1851, there had been a constant battle—sometimes three or four times a year—to keep possession. It was a battle with all the other tribes that surrounded the great tongue and Powder River basin, the richest game country in the whole world.[9]

The U.S. Congress facilitated a chain of events to maintain control of the richest game country, the Sandhills, and the territorial exchange was not easily won. In the years that followed, the Sandhills would resist permanent arrangements in two ways: through a series of droughts, and through the difficulty of drawing a line in the sand.

The decades following the Kansas-Nebraska Act are only relevant insofar as they are marked by unexpected resistance. Conflict over hunting grounds was a nuisance overcome by dividing lines and consolidating cattle ranching. But the Sandhills landscape proved more resistant to linear land claims and land patents troubling the processes of assimilation and efforts to "repeople" the territory.[10] Varying landscape patterns and processes of permanent settlement reveal the friction, as claims progress along the fringes rather than the interior of the Sandhills. In relation to forest practices, each social struggle cannot be decoupled from the landscape. Rather, the confinement of the "Indians" and the establishment of political boundaries for Euro-Americans are coupled with the carving of the Sandhills. The challenge of such a coupling undoubtedly means that more political influence will name, label, and tag the complexity of the coupling a "problem," and with every simplified problem statement comes an equally abridged "solution."

Not surprisingly, a tree-planting solution emerged through the struggle, backed by the authority of new patent agencies, land offices, and alienation acts. Federal policy created a powerful myth of tree planting to outline ownership and erase all traces of Paleo-Indian culture. The Sandhills might be converted to forest rather than agriculture, an impulse that operates from the premise that tree planting is always an act of conservation. The Sandhills are an anomaly, a micro-regionalized annex of the prairie. Other landscapes across the prairies take well to cereal grain cultivation but tend to resist tree planting. The paradox is that the Sandhills do not take well to cultivation but they do take well to tree planting.

The institution of Arbor Day is significant in relation to Nebraska's problem of occupancy and solution of tree planting. It was Nebraska politician and pomologist J. Sterling Morton who conceived of Arbor Day and persuaded fellow newspapermen to found Nebraska's first horticultural society in 1869, instituting a crusade of imported plants. His vision to declare an official day of tree planting was brought to the Nebraska Board of Agriculture and approved in 1872.[11] Arbor Day helped conjure centralized and decentralized influence simultaneously, in order to popularize tree-planting statistics rather than celebrate lasting relationships. Arbor Day became a legitimate festival and celebration, fostering widespread appeal. Its popularity streamlined planting and created an event on par with sporting competitions, using Morton's tabloid empire to circulate impressive statistics. It was not long before each state adopted the tradition, which activated another inscription of the tree unit in the American mindset.

The national forest is not only the first hand-planted forest; it is also the largest, and it lies in the heart of the Sandhills. It is unlike any other forest. It is a plantation of uniformly aged trees, generally planted too close to one another to support any other biotic affiliations, except perhaps infestation. It is a physical record of reduced species lists substantiated by tree units, a monoculture embossed across the land. But it was planted with great care and attention, the kind of consideration between plant and human life that makes it a "success" but also makes it very hard to replicate, explained in the next section. The first "artificial" national forest was an afforestation project that proved the *possibilities* of growing trees slowly over decades, a project of persistence that begins in the Sandhills.

THE NEBRASKA SANDHILLS

The Sandhills delineate the terrain across Nebraska. Within the regulated agri-surface of the state, the Sandhills ripple with dunes hemmed in by a range of bluffs, breaking the regularity of the gridded plots that surround them. The Sandhills region is awash in these kinds of contrasts:

physical cohesiveness and cultural complexity, unscathed prairie and artificial forests, expansive dryness and sizeable water resources. The plant associations echo its distinctive margins, as grass-fixed sand dunes vary from green to brown to a glowing orange depending on periods of drought. Dunes are made choppy by sporadic wind erosion, relatively low precipitation, and temperature extremes where exposed elevations have thinner surface transitions and support drought-resistant plants such as prairie sandreed (*Calamovilfa longifolia*), sandhill muhly (*Muhlenbergia pungens*), and hairy grama (*Bouteloua hirsuta*). Prominent stands of soapweed (*Yucca glauca*) iterate sporadically, while the blowout beardtongue (*Penstemon haydenii*) abbreviate steep inclines. The encounter with grassland dune plants in the Sandhills is one of endless transitions. Each manifold transition is expressed topographically and temporally.

The rawness of this dryland will not be erased or rubbed smooth by settlement. The shifting dunes are one cause, but it is the totality of the entire formation that undulates as an expression of its substrate, a major recharge zone for the Ogallala Aquifer. The Sandhills are simply the earthly expression of this buried aquatic resource. On the surface, severe surface temperatures delineate and shift the sandy soils, demarcating shallow water reserves and sharp drainage. The Sandhills express the movement of the aquifer, and plant life expresses the connection between the deep moisture and the atmosphere, blanketing the surface and migrating with the same pulses. The primary cover are grama grasses (*Bouteloua* sp.), capturing the extents that rise across the state of Nebraska in elevation from 1,880 feet to 3,600 feet in an east–west direction.[12] It feels like a sand sea, whereby the submerged, concealed thickness of the landscape provides ideal conditions for plant life because sand creates sharp drainage and freedom of movement for diverse root systems. The landscape of the Sandhills is not to be confused with desert dunes; rather, the topography undulates with verdant grasses and occasional clumps of trees. On an aerial map, the Sandhills are green, not brown.

The topographic shifts are fertile grounds for ranges of grazing ungulates but less than ideal for the standardized cultivation of land claims. A report from 1895 summarizes Sandhill formation with particular attention to the unique quality of the rhizosphere.

The Sandhills change their configuration constantly. Whenever the sand is not held together by roots of plants or by moisture, or is not otherwise protected, it is little by little carried away by wind. If a spot on a dry hill becomes bare the loose sand is blown away, a small hollow is made, the surrounding grass dies from drought, the dry sand, no longer held together by the roots, slides down into the hollow and in its turn is borne away, and thus the hollow becomes gradually larger and larger. Such "blowouts" were seen 100 meters in diameter and 15–20 meters deep. It sometimes happens that settlers a few years after breaking their land find a field transformed into a big blowout.[13]

The report goes on to explain how certain grasses can colonize "blowouts": multiplying fibrous root systems that spread differently than the tubular xylem development of woody plants—the difference between planting trees and planting grasses. The role of grasses in the formation of blowouts is anticipated a few pages later: "Here would be no stability whatever were it not for certain grasses that seem to thrive best just in these blowouts. When well established their roots bind the sand together and their decaying parts enrich the soil."[14] Such a sensitive and clear explanation of the general trend that ensues from settlement is careful to also elucidate the resultant effect on the animation of grasses. As a result, this early report promotes tough extant species such as the aptly named blowout grass (*Redfieldia flexuosa*) and lovegrass (*Eragrostis tenuis*) to help structure the anomaly of the Sandhills. But the suggested grasses were neither popular edible plants nor able to provide woody resources for the newly formed U.S. Forest Service.

Moreover, they grew on their own, sowing seeds on the wind and forming dense mesh-like formations of underground stems. These complex arrangements are particular to dormancy and the ensuing endurance to drought because plant life evolved with foraging, grazing, and nomadic relations.[15] Grasses, forbs, and sedges are the plants of the Sandhills. Despite the perceptive recommendations of early explorers, the extant rhizomatous perennial grasses did not satisfy policy—especially policy designed by forestry administrators. The U.S. Forest Service needed proof that afforestation was possible, ideally the evidence of charts, tallies, and statistics required to pass policy in congress. The cultivation of the Sandhills is a case within a case; an archetype of afforestation policy fuses with conservation to pacify existing land-based relationships.

The Nebraska Sandhills were an obvious choice because the state already established a history with the Forest Service, driven by the success of Arbor Day. Moreover, the Sandhills remained contested territory between the Pawnee, the Sioux, and the settlers.[16] At the turn of the century, land claims only scattered the margins of the Sandhills because the wind-swept interior was so difficult to humanize. But there was another, perhaps more significant reason for the U.S. Forest Service to turn to the Sandhills as a model of afforestation.

The Bessey Nursery (1902) is the first federal tree nursery, located deep in the interior of the Sandhills. The nursery is adjacent to the national forest, in Halsey, Nebraska. In relation to the Prairie States Forestry Project, it presented decades of afforestation advice to New Deal conservation, proof that planting the American drylands was an achievable mandate.[17] Beyond evidence, the nursery provided supply for the project. The techniques developed at Halsey included seed collections, propagation efforts, and transplanting failures. Each procedure charts the emergence of typology, a replicable standard necessary to design at a national scale. Typology enables professionals to normalize the specifics of ecology.

In this case, foresters newly engaged in afforestation might create a typology of tree planting, designing an inventory of tree populations and adopting tree ordinances, spacing recommendations, and tree management programs based on the Sandhills model. Typologies were reduced to the shelterbelt standard, such as wide block plantings at regular intervals and linear windbreaks. Each typology charts a very particular achievement between human and plant life. Because plants are dynamic, "success" in one territory does not always mean triumph in another. Despite the specifications of typology, plant life is not replicable. For instance, the Bessey plants prospered in the Sandhills because of its irregularity, specific to the shallow water table, sharp drainage of the sandy soils, and levels of care provided by early nurserymen. Nonetheless, the Bessey Nursery provided proof that tree survival in the prairies was a promising enterprise. Asserting typology recast the Sandhills anomaly as a prairie-state condition. Typology is the means and afforestation is the method.

The nursery is named for Charles E. Bessey (1845–1915), a professor of botany, the state botanist, and the head of the Forestry Commission of the Nebraska State Horticultural Society. His primary research theorized that the Sandhills had once been a forest.[18] Bessey's thesis advanced by subsuming tree-planting efforts among farmsteads in the Sandhills, notably the few successful claims from the Timber Culture Act. By accumulating evidence of tree planting and success of tree growth, he set the course for tree planting as an animating endeavor, one that was crucial to the occupation and settlement of the prairie states. Afforestation created an opportunity to secure more federally owned land in the fecund Sandhills. This is how, under Bessey's provocation, a landscape anomaly became a model for planting forests along the entire 100th meridian. Due to the growth of the nursery and its importance in establishing physical verification that could yield lists of tolerant tree species, this is also how the first artificial national forest developed. The appropriation of the nursery's success explains how the production of scale is implicated in the politics of plant life. Plants that take to replicable standards and perform in rows are desirable for both predictability and yield—cells on the professional index. *Index* is central to any agenda that assembles coherence across vastly differentiated landscapes.

In communities familiar with the Sandhills, the hope and frustration of life are not asserted but engendered. Federal funding and policy mandates cannot simply disperse without frustration. We can look back and identify a number of individual relationships that disrupt the larger picture; each authentic encounter relinquishes the generalizations of typology. It is not a radical conclusion that firsthand experience and long-term care engender plant growth and progress between species. What is radical is how personal investment is dismantled for reassembly as a procedure of exploitation.

FEDERAL NURSERY SUPPLY

The careful records of Charles Anderson Scott, a teamster and cook for the Nebraska Sandhills Reconnaissance Survey, afford an intimate explanation for the success of the Bessey Nursery. Scott took part in the first recorded survey of the Sandhills in 1901. In his practical, prescientific

account, a postscientific practice emerges. Two excerpts from Scott's journal reveal the difference in the valuation between "success" and "failure" and explain how his careful attention to
plant life was appropriated to generate evidence for the Prairie States Forestry Project.

> While eating out on the "Nemo Burn," one nice warm day, Miller asked me how I would
> proceed with the job, if it were given to me, to reforest the entire tract of several thou-
> sand acres. I had a plan quite well in mind and outlined a method by which pine seed
> could be sown on freshly fallen snow, from horseback. That was before the days of planes.
> I believe the plan was reported down the line to Secretary of Agriculture, Wilson, for
> adoption. Four years later I was sent back to the Black Hills to secure a ton of pine seed
> for experimental seeding of the Burn. It was sown the following year and a wonderfully
> firm stand of seedlings was secured.
>
> The first trees planted on the Dismal River National Forest were planted in the Spring
> of 1903. This planting consisted of some 5 or 6 thousand wild Ponderosa pine seedlings,
> gathered in the Black Hills National Forest in the fall of 1902, shipped to Halsey and
> heeled in over winter. When taken from the heelin-bed these seedlings appeared fresh
> and vigorous but showed signs of failure soon after they were planted in the hills. Fewer
> than 10 percent of them survived the first summer.[19]

The relationship between the moist snowy soils, the fecundity of undeveloped seed spread
at irregular intervals, and the likelihood that the horse's hooves thrust the seed into the earth
at different depths all accumulate in relation and create the context or conditions for growth.
Temporality is revealed as significant to the role of plant life in human affairs. Time is also what
is disregarded in the urge to generate replicable typologies, as "success" stories are skewed to
advantage. The narrative that "5 or 6 thousand seedlings" were planted is enough of a success
regardless of survival. Of note is how relationships are excluded from typology, a fundamental
measure of standardization. According to his diaries, Scott's experimental seeding method was
never replicated. Rather, his failed efforts to transplant seedlings persevered. Federal foresters,
eager to shape typology, simplified the necessary time it takes to work the landscape and grow
plants. As it turns out, the evidence of planting units was more important than the experience
of growing trees.

Scott's diaries offer a tender narrative that details his physical assimilation to the Sandhills
through each account of broken kettles, frisky mules, early frosts, and stomach cramps, descriptions that are only paralleled by the urge to prove himself a worthy botanist capable of identifying trees and counting annual growth rings. His efforts can be read as a collaboration between
human and plant life. Collaboration inspires difference and invites dynamic relationships that

only assume common goals of one another. From the perspective of plant life, collaboration is not a simple sharing of information. The nuance of care in a relationship is evidence of working *with* plants, with their behavior. Scott is collaborating with plants to help them grow and prosper—not on the basis of generalization to construct more projects but in the field of inquiry and attention to the living world around him.

During the first summer of the expedition, Scott took special notice of ponderosa pine (*Pinus ponderosa*) and red cedar (*Juniperus virginiana*) growing along riverbanks. This is particularly significant because these species had the lowest survival rates during the years of the Timber Culture Act, which mostly shipped seedlings from the arborized East Coast.[20] The report from Scott's expedition successfully delineated where trees were growing and outlined each acre as potential federal land reserve. Once the abundance of each species was confirmed, lands were set aside in the particularly wooded Black Hills, the current site of the Black Hills National Forest in South Dakota, now conspicuously the region of the Keystone XL pipeline extension.

At the time, there was no technical knowledge of how to care for trees, so identifying plants in the region did not pass unnoticed to federal foresters. Scott was promoted because he had intimate acquaintance with the particular textures of the region—including the messy struggles of trekking—rather than with the systematic measures of forestry. His notes describing the difficulty of the assignment are notable; he recognized that tree planting in the Sandhills was the first project of its kind ever attempted in the United States, such that "no one at the Bureau of Forestry could advise us, and the commercial nursery men of the country had no experience with this type of work and we were told we would have to use our own judgment and do the best we could."[21] The founding lines of afforestation are found in the respective attempts that necessitate practical knowledge of tree planting, a personal story of dedication and care.

Following the survey, Scott was rehired to locate a desirable nursery site and return to the Black Hills to collect seed, as shocks of cones seemed endlessly available and easily transported. The Black Hills were the supply for the Bessey plantation, as the mobile stock of plant life predisposed pines to a particular type of economy—the porous boundary between the wild and the cultivated. Scott's assignment was twofold: develop a tree-planting project in the Nebraska Sandhills, and do so by establishing a productive nursery.

The procurement of stock is a necessary first stage of every afforestation project. In the case of early American afforestation, a "successful" nursery proved to be a turning point in the mandate to build evidence and value. Prior to this, the first attempts to procure stock for planting in the prairies relied on distribution by synthetically extending plant ranges, namely, shipping plants across the continent from the East Coast. Most of these attempts ended in failure, as plants that were moved quickly could not adapt slowly to the change in climate. For the most part, this meant that expanded ranges were not a form of cultivation that could be standard-ized. But some species took off on their own, exemplified by the prolific "success" of Osage orange (*Maclura pomifera*), which is endemic to the northern plains of Texas and southern Oklahoma. The plant was spread across the country by pioneer settlers because it was easy to transplant and was excellent timber for building fence posts.[22] *Maclura* is also a prolific root-sprouting species that makes its way across overgrazed grasslands, abandoned farmland, riverine edges, and disturbed sites.

In the prairie states, *Maclura* confirmed settlement boundaries because it naturalized easily and formed large clumps that effectively helped demarcate ownership. It collaborated so well with expansionists that by the 1850s, *Maclura* was reported in New York, Delaware, Illinois, and Iowa just as it had found its way into emerging arboretum and botanical garden collections in the Northeast.[23] That is quite a distance from Texas, a distance bridged principally by humans. Miles of *Maclura* were also planted during the Prairie States Forestry Project because of its quick adaptability to a variety of conditions, which explains why it now finds its way onto plant fact sheets where it is referred to as a weed, forever invading the prairie states.[24] *Maclura* exem-plifies the changes in environmental policy, and the confusion that arises when a plant that is listed as native in one state emerges as an invader in another state. *Maclura* moves from a "suc-cess" to a "failure" as the terms of conservation shift.

Before examining the nursery for its role as primary supplier for the Prairie States Forestry Project, it is worth considering the setbacks that circulated for years between Scott and his plants. By the end of 1902, the first seedbed plantation was staked within the nursery, a plot of 136 x 136 feet (40 m) that remains entirely obscured by the subsequent rounds of cultivation that leveled dunes as the nursery expanded. This modest first site was surrounded by a five-foot wooden picket fence and covered with a four-foot picket fence because of the number of deer still roaming the region. Scott describes how seedlings did not "fare well," explaining that they were often too young or too small. In one of his earliest exchanges, he shows par-ticular kindness for a disease of the root that decimated an entire crop. In reference to the plants, he describes how "they continued to die until their little stems became fibrous enough to resist the disease, and then the little plants began to droop."[25] The disease Scott is lamenting is colloquially referred to as "dampening off," which arises during germination and is caused by soil-borne fungal pathogens such as *Fusarium, Rhizoctonia,* and *Phytophthora,* the genus also responsible for the legendary potato famines. At the time, these pathogens were undiscov-ered because the soil was not treated as a living organism by scientific endeavors until John Weaver and the grassland scientists unearthed the concealed landscape of the grasslands. The soil that Scott lamented was actually part of the ongoing modification between prairie plants and their above- and belowground environments, which operate more territorially below than above ground.

The account frames the second attempt to procure stock for the Prairie States Forestry Project, elucidated by the "successful" seedling supplied by the Bessey Nursery. Scott's reports of "success" were mobilized by omitting a decade of failures leading up to some growth, a common plight to adding nutrients slowly to the unique soils of the Sandhills. The discrepancy between his diary and federal reports of the time explicates through omission. For instance, the diary explains how a three-man survey crew was sent deep into the Sandhills to locate flat land that could serve as additional nursery grounds. He notes that because of carrying and moving so much equipment—including stakes, shovels, and packs—they had a tough time climbing. At the height that the exhausted team set down, he claims to have climbed too far after witnessing a heavy white frost one evening that "killed all the tender vegetation," urging the team to relocate one and a half miles downslope. Paying close attention to the transactions between the plant and the climate outlines the value of the slow accumulation of firsthand experience. After waiting a few days to observe the action of frost again, he decided to remain, and he finally staked out the final nursery outlines, detailed at eighty square acres. The next few years of his diaries detail each attempt to "scale up" the plantation until they were locally producing more than 125,000 trees each year.[26] Yet only the elaboration of 125,000 tree units per year materializes in the federal reports—not his exertions or the time it took him to survey and adjust accordingly.

In this version of events, "success" stories subjugate details, abstracting relationships and standardizing context. The relations between human and plant life have no economic value and cannot be replicated easily, cannot be reduced to a specification of typology. The watchfulness of Scott and the origins of the plant material from local sources are left out of the statistics that regard the Bessey Nursery a "success," just as failures are reported but not included in the shelterbelt typology.

> The results of the 1903 planting were very discouraging, but we still had faith in the project. We concluded that we had to improve on our technique of planting. We had to devise some practical method of protecting the roots of the seedlings from exposure to the air for even an instant. We concluded that the roots must be puddled rather than being kept moist in water.[27]

Scott reports that the 1904 and 1905 crops also failed. This gives rise to Scott's unique remarks on the challenges of planting in straight lines. Furrows frustrated Scott, who did not encourage straight lines. But the furrow, essentially a straight cut or trench made by a plow, was the first application of mechanization to tame the soils of the Sandhills. Scott was forced to adapt to furrow groundwork and experimented with planting seeds in an orderly manner. Yet seed survival was not contingent on linear cultivation. As cones and seedlings were shipped by freight to Halsey, approximately three hundred miles southeast from the Black Hills, Scott notes that "less than 10 percent survived" or "probably fewer than 100 survived," which expresses the difficulty of transportation. Low seed survival rates catalyze the next episode in the territory of afforestation, as the "success" of planting in the Sandhills underscores the goals of overcoming drought and addressing the socioeconomic collapse of the Dust Bowl years.

NEW DEAL CONSERVATION

1933 Passed. New Deal policies
1933 Passed. Civilian Conservation Corps program
1933 Passed. A National Plan for American Forestry
1934 Passed. The Prairie States Forestry Project
1935 Passed. The Soil Conservation Act
1938 Published. *Soils and Men*

Environmental disturbance in drylands is made visible through the combined forces of soil erosion and drought. Because erosion manifests within the surface of the territory, it tends to disrupt human life, although erosion is often exacerbated by settlement. The difference between episodic and chronic disturbance is similar to the difference between erosion and drought. Given the response to the Dust Bowl episode, and the impact of New Deal policies, the differences are significant.

Chronic stress is constant, ebbing in slow adaptive responses that have a lasting influence on regional ecology. The layers of succession demonstrated by disturbance ecology remind us that chronic stress is long-lived, while episodic stress produces dramatic, short-lived effects that dissipate over time (see Figure 3).[28] For instance, dust storms are episodic, in much the same way that sea level rise is chronic, and hurricanes are episodic. Episodic stress is easier to exploit, because it has immediate material, economic outcomes that can be measured and evaluated. Singular improvements and solutions are well suited to episodic stress. Significantly, episodes are synonymous with an event rather than a condition. Chronic stress is a disturbance that is so inadequately calibrated with human lifetimes that it cannot be managed, controlled, or tangibly reduced through a design typology. Because plant life progresses slowly compared to human lifespans, endemic or indigenous plants rarely survive chronic stress, but they are comparatively used to episodic stresses that tend to pattern their milieu: lightning, fire, flood. The grassland formation evolved with the episodic influence of dryness and drought but is comparatively less adaptable to the cycles of erosion that American expansionists produced in the prairies—as furrows, ditches, and plow incisions amalgamate as chronic disturbances.

It was through the thick, obscured film of billowing dust and airborne soil that New Deal policies, public work projects, and financial reforms emerged. The Dust Bowl years were not "natural," although they progressed through the mandates and reforms of American conservation. By contrast, the Dust Bowl exemplifies the chronic abuse of land, displacing underprivileged populations with no private property to reclaim and no wages to declare while facing escalating costs for food and shelter at a scale unseen. The response increased a sentiment of *landlessness* because federal specialists blamed the environment. Significantly, the stigmatization of those at a loss, and the rise of "problem" and "solution" policy, is mind-lullingly misguided because the "problem" was produced by a poverty of techniques that engender a distance from the environment. Humans invented the problem of the Dust Bowl. Moreover, this misinterpretation is significant because the methods, techniques, and institutional aspects of the New Deal still inform other, more global conservation policies from the Green Revolution to the Green New Deal.[29] By classifying the soil as the "problem" and conservation as the "solution," New Deal coalitions effectively normalized further environmental abuse at the national level.

Disaster Subgroup	Definition	Disaster Main Types
Geophysical	Events originating from solid earth	Earthquake, volcano, mass movement (dry)
Meteorological	Events caused by short-lived/small to meso-scale atmospheric processes (in the spectrum from minutes to days)	Storm
Hydrological	Events caused by deviations in the normal water cycle and/or overflow of bodies of water caused by wind setup	Flood, mass movement (wet)
Climatological	Events caused by long-lived/meso- to macro-scale processes (in the spectrum from intraseasonal to multidecadal climate variability)	Extreme temperature, drought, wildfire
Biological	Disaster caused by the exposure of living organisms to germs and toxic substances	Epidemic, insect infestation, animal stampede

FIGURE 3. Dryland disaster subgroup definition and classification. Episodic disturbances such as sandstorms are regular climatological processes across dryland biomes. The Dust Bowl was neither natural nor regular since it was an anthropogenic, or entirely man-made, disaster. Data from Emergency Events Database (EM-DAT). Figure redrawn after N. J. Middleton and T. Sternberg, "Climate Hazards in Drylands: A Review," *Earth-Science Reviews* 126 (November 2013): 55.

New Deal conservation forever altered the material and cultural associations of the American landscape by interlacing public concern for both natural and human resources. Roosevelt skillfully broadened conservationist models and environmentalists' warnings with the support of foresters by incorporating a concern for public health. The Dust Bowl experiences accumulate in this context of forestry-penetrating policy while small-hold farms were clinging to the surface of the newly plowed earth. As each dust storm progressed across the open landscape, it entangled gusts of wind with the torn strata of grassland soils. The Dust Bowl years recomposed the landscape as scarcity of rain intensified each mobile and persistent event. The Dust Bowl proved impossible to pin down, difficult to predict, and tough to chart since the storms were multiple but seemed to come up fast and in a very brief time blackened daylight.[30] Dust became synonymous with fear, to be avoided and curtailed. It settled slowly, then drifted quickly, staying in place by exception. That topsoil was spinning into dust stalled policy—communities do not tend to legislate a natural resource that cannot be perceived.

It was challenging to mobilize public imagination around dust: a dirty, depleted substance that embodied instability. Because the soil was invisible to the public, visible public works undertakings prevailed. This drive for reorganization pushed forestry into special interest politics, providing the administrative support for the conservation ideology set up by Pinchot. In particular, forestry expanded nationally through an accumulation of work projects accomplished by the newly established Civilian Conservation Corps (CCC).[31] As a case in point, the Forest Service supervised the conservation work of 85 percent of all CCC camps, synchronizing unemployment and conservation by operationalizing tree planting.[32] According to historian Neil Maher, conservation efforts were oriented toward planting and reforestation, overlooking the slippery exchange whereby the Forest Service promoted afforestation. Acting under the Department of Agriculture, the profession of forestry became the tree-planting agency of record.

Soil became a matter of national security; thus, it required the allied tactics of *artifact* and *index* in order to turn the amalgam of microorganisms, fungi, and living matter into a resource that could be measured, evaluated, and protected. The soil was most certainly not a rhizomatically charged organism that developed through negotiation with plant life. It was not alive and it was only vital as it pertained to crop yield and successful westward expansion. Soil was property: plotted, charted, and bound in the urge to control territory and establish power. Here, forestry and conservation truly find common ground, as trees—defined by units—are promoted as the solution to the federal problem of the "soil." Trees could be counted and tallied, extending the legacy of the fixed object, as lists of appropriate species, numbers, acres, and accounting support each New Deal conservation protocol. Through the lens of the Prairie States Forestry Project, tree units held value regardless of context, a rationale for advancing afforestation. Specifically, the project incentivized federal oversight as professional foresters took control of resources rather than relying on individual settlers to plant trees and make claims. The project of afforestation altered how private territory was claimed by federal interests, a structure designed to eliminate lands and cultures deemed unproductive via biotic erasure.

> Farmers own about 127 million acres of forest land, much of which can be greatly improved by planting. They own much land, which has been abandoned for agricultural crops, but which can be profitably utilized if planted to trees. Such a venture is possible to the farmer only if he can secure the right trees of the right size, in quantity, at low cost. He could not find such opportunities in the open market.[33]

The smooth transition describing the clearing of forest land is appropriately resolved by annexing the supply chain of tree units to a federal supply. In other words, because relations between small-hold farmers and the government did not get off to a good start during the Timber Culture Act, increasing demand for tree units would have to be more carefully managed to instill trust in the project.

Thus far, conservation emerges as a sweeping term, broadly applied to "protection" across the entire country. The conservation that took place on the western plains transpired with more difficulty, since not only was the land tougher to settle, but the cultural transition was ultimately more stubborn. Nebraska is a case in point and provides a more detailed reading of the hundreds of thousands of micro-conflicts unfolding as a result of early conservation protocols. At the time, the term *Indian* was applied to all Native Americans encountered by expansionists, following early European explorers.[34] New Deal conservation continued to deny the 13,500-year-old history of indigenous peoples and the evidence that areas of the Plains were settled by Paleo-Indians before 5,000 B.C.[35] It also denies that planting is an inherently social— not ecological—process. There is nothing ecological about planting trees across the grasslands and sandhills of Nebraska. Rather, there is the process of turning temporal patterns into bounded lots of private property.

The attempt to make conservation easy made it trivial. The federal government set up the conditions for mainstreaming exploitation by manipulating ecological terminology and passing it through the benign qualities of conservation, not unlike the progression of turning industrially induced drought into a desertification crisis to gain territory. To reiterate remarks from the introduction, this is not a historical anecdote; it is a carefully appropriated tradition that relies on singular solutions like tree planting to distract from the difficulty of changing course, of addressing environmental abuse.

THE POSSIBILITIES OF TYPOLOGY

The records kept by the Bessey Nursery substantiated the possibilities of afforestation, providing immediate confirmation for the Prairie States Forestry Project under New Deal conservation. Thus, the traditions of afforestation originate in the appropriation of a twenty-five-year tree-planting effort, led by Charles Anderson Scott in the Nebraska Sandhills, and possibilities of his stewardship.

> During the years that I was in charge of the project we had visits from a goodly number of the Bureau Officials, including Mr. Gifford Pinchot, Chief Forester, who spent 2 days with us. He was much impressed with the possibilities of the project and the progress we had made up to the time of his visit. He assured me of his faith in the project and in our ability to carry it out.[36]

Each careful accomplishment is put into fast-forward as it figures prominently in the literature that surrounds the popularization of the Prairies States Forestry Project, notably, *Possibilities of Shelterbelt Planting in the Plains Region,* a widely distributed attempt at logistics. Part edited volume, part manual, part master plan, the publication identifies multiple arrangements of professional expertise and explores different perspectives on the formation, climate, geology, vegetation, and hydrology of the prairie states. The publication is neatly divided into distinctive modes of practice that include references to the planting success at Bessey Nursery, without mention of Scott and without an acknowledgment of his personal and practical field tests. It also carefully omits the extensive efforts of grassland scientists to elevate the role of relationships between plants in the region and justify treelessness.

The domestication of each "success" story raises the question of how scale advances through substantiation in linear questions. Linear scale has little significance to the magnitude of plant life. Scott's journal provides an entry into a model of operations that resists scale and embeds progressive change, greater detail, and informal gradation inspired by the study of individual plant behaviors. As a result, attention paid to plant life embeds a different type of knowledge, one that is not built upon dominant or climax species, nor on simplified models that only find value in units—worth noting because at this point the procedures of *artifact* are so brutally hardened into the culture that any attention to *life* in relation to plants is simply disregarded. Index advances typology, but only insofar as plants were already stripped of their aliveness.

To replicate is to repeat, an efficient convention valued for the quick pace of duplication processes that depend on removing complications. In design fields, replicability is inherent to the use of typology, an ordered and reliable set of types inherited through archaeology into architecture. For instance, a typology of building structures would include theaters, arches, columns, baths, and aqueducts. Context is always absent from typology. Instead, techniques of construction and installation are standardized by the metrics of analysis that rely on the most visible and static forms of knowledge. In forestation projects, typology manifests as the rows, belts, and grids that continue to define global planting and propagation projects. The specificity of plant life receives little attention in these accounts, just as relationships and time might complicate the outcomes. Typology replicates outdated practices and is refutable in botanical or ecological terms despite the fact that shelterbelt planting continues to repeat across the planet. In other words, the continued use of typology does not align with contemporary knowledge, indicating in greater detail how deeply colonial-era assumptions are embedded in scientific literature and design practices.

The manual *Possibilities of Shelterbelt Planting in the Plains Region* is replete with diagrams that chart the *possibilities* demonstrating the potential of the project by subtly elevating the Sandhills as a "success" while subverting its unique landscape within the region. In one stand-out chart, the use of "survey strips" links typology with the tradition of pilot projects. Each two-mile survey strip accumulates evidence of past plantings by taking a *representative sample* in an east–west line across each state. Each line, or transect, traverses the land as a means to average the accumulation of remnant shelterbelts and build evidence of where plantings survived.[37] The most promising accumulation is in Nebraska, simply due to the fact that it includes Sandhills afforestation, or the Nebraska national forest. Each stratum of evidence depends on the confidence of standardizing the efforts of early tree planting in order to promote the Prairie States Forestry Project. Selecting a particular transect is a worthwhile strategy because it provides evidence by cutting into land acquisition programs that have already planted trees. But it is much more ambiguous to insist on typology within the delineations of local efforts and the history of indigenous practices. This is where scale emerges as a particularly problematic device, as typology establishes a prevailing system that is superimposed on the ground. Moreover, because the tree is the symbol of the project—a tactic inherited from the success of Arbor Day—the project taps into the environmental consciousness because there is nothing unnatural about a tree.

> In 1926, 34,000 trees were distributed to 105 farmers in 48 counties, and 53 percent of the trees lived. In 1927, 186,000 were supplied to 1,161 farmers in 992 counties; 700,000 were furnished in 1928 and a similar number in 1929. The activity has kept on increasing until, in 1934, the distribution was 1,125,000 trees.[38]

Possibilities of Shelterbelt Planting in the Plains Region was approved in 1934, the same year that Company 798 of the CCC was moved from Fort Robinson, Nebraska, to the Bessey District.

The claim is also made that "1934" marks the beginning of the Dust Bowl years. This important year sped up national connections established in 1933, namely between soil conservation and afforestation. The common unit—the tree—holds together the collective, believing in planting as an organized conservation model, when it was actually a New Deal incentive to produce social industry. *Possibilities of Shelterbelt Planting in the Plains Region* accelerated the ambitions of forestry and quickened the pace of politics, offering a historical marker from which to consider how sometimes the most creative human acts do not acknowledge the depth of their environmental incongruity.

The typologies of afforestation first appear in the shelterbelt manual as a commitment to conservation policy, with three distinct spatial outcomes, outlined in the remainder of this chapter.

The Field Shelterbelt

Land should be acquired 10 rods wide and a planted strip made at 8 rods wide. It is estimated that 1,282,120 acres should be planted.

A rod is approximately equal to five meters and serves as the unit of measurement for surveyors, whereby a stake is inserted at every *rod*. This typology stipulates the acquisition of a fifty-meter swath of land whereby a forty-meter belt is planted, allowing for some give on either side for the spread of roots and crowns. The setback also enables the belt to be well fenced from grazing livestock. Although the manual acknowledges that it is unfeasible to expect a continuous row across the continent, to be fully effective it suggests that intervals of field shelterbelts are no greater than a quarter mile to a mile. The field shelterbelt is the most common shelterbelt typology and is described to be roof-like in section, whereby the species selection allows for the tallest trees in the center, edged successively by shorter trees and tall shrubs (see Figure 4).

Farmstead Windbreaks

Where the need is great, the plan should be to aid farmers in establishing windbreaks around farmsteads. For this purpose, 897,880 acres should be planted.

The farmstead typology stipulates that agreements should be made through cooperative contracts between private landowners and the state. To qualify, assessment was based on soil types that are deemed to be more difficult soils, a typology to aid farmers in establishing windbreaks around farmsteads. Although the spatial typology is exactly the same as the field shelterbelt, tree planting was instated and maintained by the CCC, ranking the farm windbreak as social

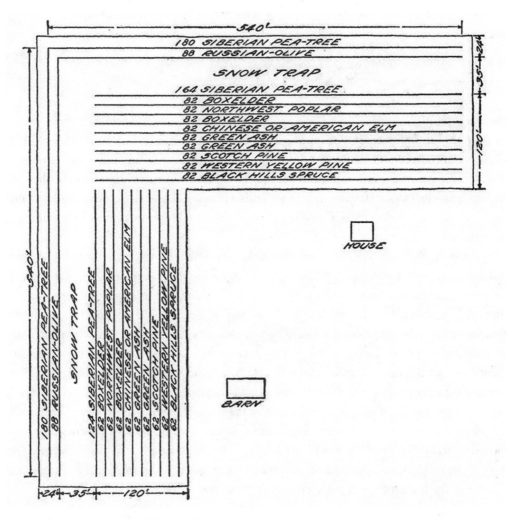

FIGURE 4. The field shelterbelt. A design typology includes recommendations for tree planting, regardless of the extant landscape. The planting plan is the ultimate afforestation standard because it omits context and enables duplication. Lake States Forest Experimentation Station, U.S. Forest Service, *Possibilities of Shelterbelt Planting in the Plains Region* (Washington, D.C.: Government Printing Office, 1935), 00.

construction. It is entirely unclear why this typology is called a windbreak, as opposed to a shelterbelt, other than as a means to proliferate a difference between planting for domestic value versus contributing to the federal project.

Block Planting

> In order to prevent further erosion 4,099,000 acres to be placed under forest and range management. 400,000 acres of this area will be planted in more or less solid blocks of trees.

The block-planting typology is unique because it falls under forest management and was established in the Sandhills region. The spacing is not specific to species; rather, it references earlier failed attempts that arrayed the trees at four-foot grids. Instead of specifying species, the manual is specific about the heavy maintenance required in thinning when the blocks are so densely planted, and advocates instead for eight-foot grids. Of note is the specific mention that many blocks are now covered in grass, thus it suggests deep-rooting trees are capable of providing a more stable protective cover, a fallacy of afforestation.

Typology is designed as a technical abstraction that excludes more than it includes. The abstraction appears neutral, but it is biotically and culturally specific, growing out of particular contexts, conflicts, and circumstances. Afforestation in Nebraska was territorially specific but culturally replicable. The cultural politics of moving species across the continent in quantity epitomizes the shift away from the emerging practices in ecology and toward the redundant horticultural origins of the eighteenth century, as expertise in bio-prospecting, plant identification, transport, and acclimatization worked hand-in-hand with European colonization. But the movements of plants across the prairies was not encouraged for the spices, medicinal drugs, perfumes, or dyestuffs of European trade. Rather, the establishment of the Bessey Nursery and the subsequent popularization of afforestation represents an ecological fallacy borne of the systematic effort to claim land. In turn we are a species that has favored those plants that grow in rows, along furrows, and along fences, in predictable patterns with foreseeable outcomes, an injustice of the insistence on typology that endures today.

6

Ulmus pumila L.

THE STORY THAT UNITES THE GENUS *Ulmacea* begins with its winged seed. The linear narrative of reproduction, release and development, tends to describe seed behavior, which obscures the reciprocity between plants and their environment. Attention to seed dispersal, mobility, and communication disrupts the linearity of prediction, as multitudinous seed swarms, lands, and burrows in decisive collaboration across timescales out of sync with unit measure. Dispersal is a safe method for dealing with highly disturbed systems, which is why drylands are the terrain of seeds, the reach of plant life beyond planting. The seeds of *Ulmacea* mature in the center of mobile papery fruit called a samara, a lightweight bundle quite unlike the hooks of burs or the cones of cycads. Winged samaras are easily picked up by prevailing winds that convey volumes across vast distances. As the samara breaks down, or is torn apart, the nestled seeds continue to tumble, grow dormant, or are urged to germinate. Seed dispersal is not linear; it persists in time and extends in space, elaborating the surface of the landscape. Across the American prairies, seeds rearrange regional ecology and settle within expanding human communities. Here, behavior extends from the living plant, into the environment through the seed, raising a striking question: Do seeds behave?

The prolific seed habit of *Ulmus pumila* sheds light on the narrative between introductions by humans and invasions by plants. Each type of botanical progress informs the theoretical idea presented here, namely how the messy, surprising aliveness of plants remains at the center of contemporary debates and the politics of afforestation. The disturbed grounds of human settlement are where plants with rapid growth, flowering, and seed production prosper. Moreover, due to millions of years of evolution, plant life excels through specialization despite the processes of valuation that include or exclude to decipher singular solutions. In other words, some plants thrive with humans, but we tend to prefer those that struggle. In the matrix of global plant exchange, *Ulmus pumila* is a case in point because it advances by including human life in the production and dissemination of its species, colonizing the bare ground, slowly shading the soil and adding organic content to the developing mat under its immediate canopy, while gaining widespread territory through the pulse of its decisive seed.

Seeds are the external reproductive unit of flowering plants. One seed typically holds two tiny meristems, a dormant root and a shoot package enveloped in a nutritive tissue. This embryonic package is mobile and designed to significantly expand plant ranges. Seeds travel across the world, gaining momentum as they cling to animal fur, hide in the cuffs of farmers, ride the wind, or cycle through the water. Each journey is helped along by the elements lifted, twisted, soaked, and scarred, until it is grounded. The package is liberated by its association with the elements, as germination animates the seed package. But seeds do not differentiate germination agents; the seed is unbiased whether released by forest fire or human harvest. Perhaps if plant life were recognized as a social and collaborative force, then both industrial cultivation and global conservation would appear peculiar, positively discrediting those predominate forms of exploitation that hide behind a fog of conservation and sustainability in order to ultimately maintain the veil of human domination.

Germination is simply development from seed. While many seeds germinate in the environments, still more germinate under human influence. Selective or "forced" germination explains that seed evolution progresses along with the human preference for sedentary lifestyles and industrial agriculture. The incredible success of forced germination mirrors the achievement of human settlement, as sedentary seed-based diets and cultivation slowly replaced foraging and nomadism across the planet. Fast-forward, and only three crops (wheat, rice, maize) now account for 60 percent of the world's food.[1] This is an achievement for humankind and plant life alike. Both prosper, both thrive, both rely on one another, but only one is entirely dependent on the association. The seed persists across human timescales because humans are planting species. But it does not proceed unimpeded; seeds evolve with human life.

Domesticated seed benefits from a close relationship with planting because of a reliance on humans to extend ranges, yet human-assisted migration is rarely described as an evolutionary trait. This points to a discrepancy, consistent only with removing humans from the equation. The botanical narrative attributes dispersal to "cultivation," a noun that fittingly abstracts the close connection between seeds and humans by replacing the relationship with a generic reference to farming.[2] Seeds travel further and further afield because humans pick seed, pocket seed, and store seed. So, if long-range dispersal is actually dependent on the acts of individual humans, then why not involve the influence?

The behavior of domesticated seed is a reminder of the genetic simplification designed by humans. The relationship between animated seed and human initiative is reciprocal. Another word for the relationship is collaboration. Either way, an alliance emerges between plants and humans as certain seeds are favored over others. As seeds are carefully selected for particular attributes, isolated traits advance human settlement; perhaps it started from a craving for specific flavors, progressed for disease resistance, and was scrupulously defined by spore retention. At any scale, preserving certain genes and excluding others means that humans breed highly localized landraces, engendering seeds that are averse to extremes and variation, because cultivation favors temperate conditions and excludes extremities, such as drylands. We are selecting plants that thrive in the same conditions as humans. The persistent and unchecked effect marks diversity and regulates food supply reliant on only a few highly predictable plant species. Fast-forward and wheat, rice, and maize create a bargain written in the surpluses, famines, and pesticides the world over.

Conversely, wild seed is marked by a spontaneity animated by survival. Spontaneity is found in expansion of range, and the distance a seed might disperse beyond its immediate range leads to potential germination in different climates. Seeds help plants expand their range. The capacity to seed means that some plants achieve cycles of retreat and advance that far outpace creature movements. Each macro-migration lies in stark contrast to the human imaginary that plants are sessile, or only capable of micro-migrations. Once released from the sessile plant, seeds are mobile, independent participants. For the following pages, as seeds dominate the narrative, imagine how individual plants travel, move purposefully, and influence wide-ranging contexts.

The incredibly independent, preprogrammed, and strategic behavior of *Ulmus* seed coalesces with planting activity in the Great Plains. Here, *Ulmus* is a pioneer, riding along with a population moving rapidly into new areas, producing offspring and subverting weaker species. Seeds encourage expansion while behavior raises the topic of dormancy, a strategy that helps the plant deal with the vast number of environmental variables it meets as it expands. Plant biologist Anthony Trewavas describes dormancy as a factor that is dependent on what other seeds are actually doing.[3] Accordingly, Trewavas explains that seeds are capable of making decisions between dormancy and germination, in what he describes as an expression of memory or "molecular states." Seeds are informed by cues in the environment, and "bank" on changing inputs or take the risk of waiting it out. Remarkably, seed response is not limited to local environmental cues because seeds also draw on the embedded experience of their parents.[4] Visualize an individual seed, snug in its samara membrane. The seed is assessing the environment for indicators of potential growth or germination, at the same time as it surveys the quantities of adjacent seed, before *deciding* whether to send out and anchor its root tip or grow dormant and ride the wind.

Seed independence brings up a relevant attribute referred to botanically as *plasticity*. Plasticity is the amount that an individual plant can be modified by its environment.[5] The term speaks to the ability of plant life to alter development in relation to external cues, an essential modification because plants are sessile. When a plant senses cues in its environment, it adjusts "plastically" to the variations. Plasticity extends beyond the individual plant through its detached seed, the seed entirely removed from the adult plant. Emerging scholarship in plant behavioral science endorses seed movement and dispersal as an intriguing subset of adult plant behavior, elucidating that seeds are programmed with different periods of dormancy: "In the seed with just two meristems, the only potential for plasticity will be in its timing of germination, but this is behavior and quite complex at that."[6] Here, plasticity is described by signals received at different times, in order for the adult plant to ensure germination takes place in the most equitable environment for juvenile growth. But how is the signal sent? Once the seeds are released, can they communicate? Is wild seed behaviorally more vigorous than cultivated seed? These questions are difficult to address without paying attention to some of the most prolific seeders, those that are often called invasive plants, aggressors, and superweeds.

The study of *Ulmus* is not located in the fragments of global integration that lend currency to the exchange between plants and humans. Instead, the study of *Ulmus* is fraught in relation to either the ruins of Dutch elm disease or debates in the aggressive spread of so-called Chinese elm. Any research on *Ulmus* must now contend with a science struggling to either solve a global pandemic of devastating beetles and fungus or define the intimate intraspecies mixtures between Asiatic introductions and "native" plants. Perhaps this is reason enough to pay attention to the particularities of seed, a less-industrious project of expansion brought about by plant rather than human life.

Species Confusion

Ulmus parvifolia (Chinese elm) is a relative often confused for *Ulmus pumila* (Siberian elm). While both are considered "Asiatic elms" by the botanic community, they are different plants with different behaviors. The family *Ulmus* sp. is desired for its wood, a desirable trait that pairs with its resistance to splitting, rendering the plant an ideal timber tree. Prior to introductions, Americans sought *U. americana* L. in the American drylands, materializing as wagon wheels, ax handles, coffins, and drums. For the sake of clarity within confusion, the Latin *Ulmus pumila* L., whereby the "L." stands for Linnaeus, is set apart by its prolific seed, often associated with "invasion." What all this adds up to is that the prolific seed of *U. pumila* is overlooked by the attentions of foresters at the time, due to the influence of its resource-based attributes. *U. pumila* is invigorated by multiple relationships, taking advantage of wind and forester and expanding its range across the American drylands.

Significantly, the other difference is that *U. parvifolia* is a hard woody tree that grows up straight with few serious pests, while *U. pumila* tends to be short lived, disordered, and defiantly root sprouting. The conflation between species is an example of how the *index* of specialized units creeps into ecological contexts unnoticed. This is not to say that the intrusion is negative. Rather, it deserves notice because humans are very much a part of the environment, a partner by virtue of invasions, introductions, and exterminations. *U. pumila* thrives across the American prairie, finding early success in germination because of a new dispersal agent, expert foresters. Each catalog and species specification spreads quantities of *U. pumila* as profusely as its seed, materializing *index* as a powerful spatial device that extends far beyond a theoretical path. *U. pumila* is popularized as "Chinese" elm because it was collected in China.

U. pumila confuses the boundary between capacious habit and calculated prediction, a feature of behavior that evolved over millions of years within the context of sandy lands. Significantly, the confusion between elms sneaks its way into the American Prairie States Shelterbelt Project, as millions of seedlings were specified, germinated by hand, and planted across millions of acres. It was not until 1948 that the distinction between "Chinese" and "Siberian" plants was clarified in the publication "A Report on *Ulmus Pumila* in the Great Plains Region of the United States." The following statement was released concerning the oversight: "From a study of the history of oriental elms that have been planted in the Great Plains, the author concludes that *Ulmus pumila* should properly be called Siberian elm and *U. parvifolia*, Chinese elm."[7] The report makes the distinction clear, despite decades of intensive cultivation; accordingly, the report goes on to say:

> The "Chinese elm" has been widely distributed and used, especially in the Great Plains Region of the United States, during the past thirty years. It was the main windbreak tree in the shelterbelts planted by the U.S. Forest Service. It comprises the largest proportion of farmstead windbreaks and protection plantings made during the last two decades by farmers in that region. And observation will show that 75 percent or more of all new street, shade, and park trees planted within this area in the last few years are "Chinese elms."[8]

What happens when American pioneers collect seeds, misinterpret their origins, and make deals with overenthusiastic experts? When the species in question is well adapted to aridity and disturbance, the exchange funds short-term achievement as *U. pumila* advances, ruthlessly breaking typology. The so-called Chinese elm is celebrated and overplanted, as it takes to the degraded prairie landscape through tree-planting campaigns because it is so wildly independent.

Outside the confused speed of human introductions, *U. pumila* behaves through two distinct adaptations that pair to expand its range or influence. *U. pumila* is adapted to disturbance, from overgrazing to drought, and spreads seed in order to push into new lands. It produces prolific amounts to guarantee survival. As a result, *U. pumila* tends to inscribe the transition between grassland and desert, or semi to true desert, where rainfall drops or increases, where ecological cycles shift on the edge between biomes. But plants that profit from disturbance and spread into new lands are often associated with what conservation mandates term invasion.

The vast indigenous habit of *U. pumila* covers central Mongolia, southeastern Russia, and western China, including the Horqin Sandy Land and Otindag Sandy Land of China, regions of the Chinese Three Norths Shelter System. *U. pumila* is often referred to as the "last tree" in the transition between biomes.[9] It is a sentinel to desert dwellers, a sign of life, where resources from wood fuel to water might reside. As a wind-pollinated tree, *U. pumila* is characteristic of open, sandy regions where it grows sparsely across landscapes characterized by either single or more often small clusters of trees.[10] Because it thrives in drylands, *U. pumila* is considered a drought-tolerant plant, which explains its partiality for developing in small groups called clusters. Clustering enables plants to share limited resources, especially water. The patchy habit of clusters helps protect progeny. Using the wind as a medium, scattered clusters of *U. pumila* trees distribute seed over steppe slopes by working with the variability of seasonal currents. The adult tree gains mobility as the wind picks up, liberating the seed bank from its canopy, circulating, mingling, and transporting each unit as a vector of environmental and biotic relationships. The first scattering of its lightweight samaras sets the stage for a vast seed migration.

Across the sand land regions, samaras do not stay airborne for long. Rather, they invade the space just above and below the ground surface, in the mesosphere near the surface and the atmosphere near the ground. Seeds advance in both vertical and horizontal ranges, a thickened space that negotiates a mix of airborne and surficial particulate. This is a landscape of micro-movements where sandy elements mingle in a decisive but turbulent resettlement. The mobility of the surface is driven by the small-scale distribution of existing plants, buried rhizomes, and soil horizons, determining where the radicle decides to emerge and begin to seek water-bearing pockets.[11] The seed package is typically deposited as an assemblage of other progeny, a bundle that anticipates upward rising currents to take individuals away from the group, furthering distances. Oftentimes, distance is dependent on the condition of the samara. The patchy whirling dispersal is informed by micro-depth relations, as seed spreads locally in millimeters and samaras ride regionally across miles. As seeds burrow, the granularity of the mobile surface is conveyed to the seed through the papery samara, as it attunes to alterations in weight, depth, and burial that ultimately prompt seed germination. The inclination of wild seed suggests that the adult plant is reaching to gain new territory, a behavior highly adapted to change.

Invasion success is behavior so attuned to change that it mobilizes from a multitude of positions, securing survival among competitive forces from sandstorms to drought, from nature lovers to technocrats bent on eradication. The demarcation of *U. pumila* along the transition

between sandy land and grassland is plasticity, a sophisticated response to environmental variation. Remarkably, plasticity might also be found in the seed. Plants that grow fast often means
short lifespans, tempered by the production of copious amounts of offspring.[12] In the case of
U. pumila, prolific seed ensures successful germination on open ground, far enough from the
adult tree so as not to compete. The invasion into other biomes is not an attack; it shows a
distinguished capacity for seeds to relocate, and for roots and shoots to locate resources. Unfortunately, this dispersal is currently associated with "invasive" species. It is not surprising
that there is extensive literature to back up plasticity in stress-tolerant plants, but it is surprising that these plants are considered "invasive."[13] An "invasive" plant tends to be stress tolerant because of the twin features of rapid establishment, and resistance to environmental stress
and herbivory. The success of the plant is marked by fast growth once the seed *decides* to
germinate.

The decision for the buried seed to germinate or remain dormant is particular to a reading of plant behavior. As Trewavas prompts, dormancy is factor-dependent on the movements
and decisions of other seeds. Accordingly, seeds are doing many things at once. They are track-
ing water content of soil at the molecular level when they land and waiting to time their migration, since dispersal coincides with periods of seasonal windiness.[14] The mobile seed is at once
adapted to movement and stasis; it must travel and anchor, bury and move its airborne self.
Windstorms and exposed land are largely an overwhelming prospect for plant life, as scarcity
of resources tends to repress reproduction and thus inhibit germination. Once the seed coat
breaks, the plant begins to prioritize biomass allocation in the roots, the behavioral expression
of plant life below the surface.[15] *U. pumila* recognizes the heavily eroded American prairie soils,
as preadapted root tips meet dryland soil microorganisms.

Self-Organizing Behavior

Ulmus pumila thrives in conditions that are adverse for most trees. The prolific seed of
U. pumila endures because it is coded to settle into dryness. The fascinating question of germination sheds light on self-organizing behavior, across the full life history of plant life, tran-
scending the outline of a tree unit. Plant behavior is significant because it advances either
with the human project or against it. Extrapolate from there and seed selection comes into
sharp focus, as plants are invited into human worlds or set apart from them. Of note to the
case of *U. pumila* is that the early appreciation of rapid establishment and ease of germination
was interpreted as a positive attribute, one that could be exploited by afforestation. While
U. pumila proliferates in tandem with human introductions at first, as plants mature, they borrow local breezes and ride across the horizontal desert in gusts, flurries, and gales. Seeds are

exposed to variable periods of drought and degrees of sand burial after dispersal.[16] Accordingly, as *U. pumila* begins to self-organize across the prairie states, it spreads rapidly on its own.

The strategy for survival among self-organizers is embedded in a careful code implanted by the adult plant into seed. The study of behavior reveals that the adult plant programs different stages of dormancy into each seed to maximize endurance across a range of potential variables because it is very difficult to predict where and when seed will settle. Inside the seed, each dormant meristem carries the code, safeguarded by a protective coat. Seeds are programmed by "code" from the adult plant, and, simply translated, the code receives signals. The term *code* is another way of saying that seeds carry genes and different regulators. As Trewavas notes, "These signals have a recognizable ecological basis, they are designed to try and ensure germination takes place in what will be the most equitable environment for seedling and juvenile growth."[17] Timed dormancy suggests self-organizing behavior because dormant seeds can opt in or out of germination to reduce risk. The evolution of mobile juveniles and sedentary adults is quite unlike the reproductive strategy of humans and other creatures, who tend to produce dependency between sedentary juveniles and mobile adults. Because the adult plant remains sedentary, the juveniles are persuasively mobile, as survival between seeds is powerfully communicated by signal and preprogrammed by code. Moreover, the study of behavior reminds us that signals percolate between plants, plantlets, *and* dormant seeds.

The decision to postpone germination is coded into the seed. Theophrastus lends some currency to the question of whether or not seeds *decide* to calibrate their advance: "For though every living thing is nourished, growth will result only when it is stirred by the impetus for germination."[18] Perhaps the question of whether or not plant life exhibits behavior is found in this early use of the term *impetus* conveyed before the plant was objectified and rendered inert. Impetus resonates with *orme,* as a force of movement associated with informed compulsion, which sheds different light on what botanists now call seed migration. A closer look inside the samara-encased seed suggests that plasticity progresses at distances and depths well beyond the purview of the adult plant. Maybe seeds stir by design.

U. pumila seed settles into novel local habits seeking open ground, and the disturbance of settlement offers a range of potential sites: an aspiring Nebraska shelterbelt, a roadside ditch, a plowed field, the cracks of a foundation. Seeds are long lived and concentrate on mobility until the right conditions instigate animation. Once signals are received inside the papery coating, the radicle punctuates the seed coat and begins to participate in plant formation. Inside the seed, a soft food supply offers protein, fats, and starch to the juvenile, animated root cells. This first sign of germination is typically followed by the formation of root hairs, the fine-tuned threads of absorption and exchange that seek a combined intelligence in the rhizosphere. As the root wanders, the seed turns over and adjusts itself to establish a comfortable relationship to depth.[19] The endosperm breaks again to allow for the emergence of its aerial parts, which unfurl to reach the surface. These ongoing movements and fine calibrations establish the growing point for the seedling as it begins to seek environmental inputs. The internal intelligence of the seed is exemplified in the hidden cotyledon, a leafy-looking structure that is split in two and called a dicot.[20] The cotyledon is often informally called the seed leaf because it looks like a leaf and sometimes retains a piece of the seed, having had difficulty shedding the coating on its rise to the surface. But this delicate emergence is not a leaf; it is part of the embryo, as it lived inside the seed throughout its mobile migration. The manifestation that emerges above the ground is the aerial portion of the same root-shoot organism that is not yet a plant. Leaves develop outside the embryo and begin to photosynthesize and grow up a plant. Therefore, a cotyledon is developmentally distinct from the plant, although they are functionally similar.

The plantlet reacts internally, self-organizing with cues from the atmosphere and the rhizosphere. According to Trewavas, trees are especially good at self-organizing; he describes the tree as a complex network, in which fairly simple rules of interaction qualitatively change with size.[21] The woody tree, consisting of millions of repeating modular structures, is actually an extremely flexible structure because it can marshal large numbers of its modules toward necessary objectives, assimilating internal and external inputs to rearrange itself. This is an example of how organisms play a part in their own evolution. Imagine that the *U. pumila* seed, a future tree, is carried by the wind, collected by humans, transported across oceans, and resettled into a federal nursery bed. At each stage of the journey, it is processing dynamic inputs. Such a feat of the imagination is difficult to picture, or more specifically the management of seed would be difficult to restructure if the intelligence of plant life was more commonly accepted. Describing plant behavior is not an appeal to equate plant life with creature life. Seeds do not *feel*, but they are also not entirely inert. The seed is something other than plant life but something nonetheless significant on its own terms.

The success and abundance of self-organizing *U. pumila* seeds was a source of pride for federal foresters. Although it was often listed as "Chinese elm," a straighter, more predicable elm, specialists took the credit for its favorability to the plains and resistance to disease, a conclusion drawn from reportage between specialists.[22] One such report from a farmer in Bridgeport, Nebraska, details planting in the lexicon required for increasing units:

Trees planted on May 1, 1918, were reported upon as follows on April 7th, 1922: "Trees when received were not over 3 feet high and about the size of a lead pencil. On

November 1, 1921, by actual measurement, they were 16 to 19 inches in circumference and from 15 to 25 feet high.[23]

As a result of early accounts by farmer-developers, *U. pumila* finds itself spreading across sanctioned planting lists compiled by eager federal foresters (see Figure 5). More often than not, it is listed as "Siberian elm." The plant, along with its behaviors, is recommended in every state of the shelterbelt project from Texas to North Dakota, a responsibility not held by any other individual species.[24] While its Latin binomial was correctly identified, it was mislabeled the "Dwarf Asiatic Elm," adding to the confusion in elm culture. More significantly, it was typically listed and recommended along with *Ulmus Americana* L., the American elm, a beloved symbol of empire. The plant, superficially labeled "Siberian elm," proliferates units based on aboveground form, the quick performance of species recommendation lists. But the adaptability of *U. pumila* is located in the rhizosphere.

U. pumila is considered an azonal species, meaning it is more severely influenced by the soil than by the climate. Being azonal was an immediate advantage to a plant accustomed to disturbance and sandy lands, since the American grassland was a thick store of nutrients, rich in microorganisms and abundant in organic residues and carbon.[25] Upon its introduction to the North American grassland biome, it mobilized spread with the prevailing winds, performing, behaving, and evaluating its new sandy context. Samaras were released and found sufficient movement, as seeds prospered in clumps among the prairie grasses. The profusion of *U. pumila* was quite unlike the efforts to establish American elm trees in the prairies because they rarely survived the strong, windy winters.[26] *U. pumila* self-organizes and persists temporally through cycles of dormancy. The differences between elms were invisible to speculators.

Urged on by the creative agency of *index,* newly planted *U. pumila* spreads along moist banks, within seasonal drought, on disturbed or pasture land, and in cold winters, making it one of the most tolerant deciduous trees of the prairies. It collaborates well with humans, as it thrives by extending ranges, by migrating, moving, and persisting. *U. pumila* dominates because it evolves with human life.

Species introduced for experimental purposes will not be used on a large scale until they have been proved suitable for shelterbelt planting. Latitudinal limitations will be observed in the movement of seed from one locality to another in order to retain the adaptation a species has acquired for a particular environment. In many cases localized collection of seed and planting of trees grown from it will be advisable. It has nevertheless been found necessary to go outside the zone to secure a part of the seed, particularly of Chinese elms.[27]

Mature size and common name	Maximum height (feet)	Height growth	Useful life
Tall trees:			
Cottonwood	80	Fast_ _ _ _ _ _ _	Medium
White willow	50	_ _ _do._ _ _ _ _	do.
Siberian elm	60	_ _ _do._ _ _ _ _	do.
Sycamore	60	Medium_ _ _ _	do.
Shortleaf pine (south only)	60	_ _ _do._ _ _ _ _	do.
Ponderosa pine	60	Slow_ _ _ _ _ _	Long
Austrian pine	60	_ _ _do._ _ _ _ _	do.
Medium trees:			
Northern catalpa	40	Fast_ _ _ _ _ _ _	Medium
Black willow	40	_ _ _do._ _ _ _ _	Short
Golden willow	30	_ _ _do._ _ _ _ _	do.
Green ash	45	Medium_ _ _ _	Medium
Hackberry	45	_ _ _do._ _ _ _ _	do.
Bur oak	45	Slow_ _ _ _ _ _	Long
Scots pine	45	Medium_ _ _ _	Medium
Jack pine	45	_ _ _do._ _ _ _ _	do.
Eastern red cedar	45	Slow_ _ _ _ _ _	Long
Short trees:			
Boxelder	25	Fast_ _ _ _ _ _ _	Short
Russian-olive	25	_ _ _ _ _do._ _ _	Medium
Diamond willow	15	Medium_ _ _ _	Short
Russian mulberry	20	_ _ _ _ _do._ _ _	Medium
Osage-orange	25	_ _ _ _ _do._ _ _	Long
Rocky Mtn. cedar	25	Slow_ _ _ _ _ _	do.
Shrubs:			
American plum	10	Fast_ _ _ _ _ _ _	Medium
Chokecherry	10	_ _ _ _ _do._ _ _	do.
Tamarisk (south only)	10	_ _ _ _ _do._ _ _	Short
Purple willow	6	_ _ _ _ _do._ _ _	do.
Common lilac	8	Medium_ _ _ _	Medium
Honeysuckle (north only)	8	_ _ _ _ _do._ _ _	Short
Caragana (north only)	10	_ _ _ _ _do._ _ _	Medium

MEDIUM TO DEEP UPLAND SOILS (SILTY OR CLAYEY LOAMS)

Mature size and common name	Maximum height (feet)	Height growth	Useful life
Tall trees:			
Siberian elm	45	Fast_ _ _ _ _ _	Medium
Shortleaf pine (south only)	45	Slow_ _ _ _ _	do.
Ponderosa pine	50	_ _ _do._ _ _ _ _	Long
Austrian pine	45	_ _ _do._ _ _ _ _	Medium
Medium trees:			
Green ash	35	Medium_ _ _ _	do.
Hackberry	35	_ _ _do._ _ _ _ _	do.
Bur oak	35	Slow_ _ _ _ _	Long
Eastern red cedar	35	_ _ _do._ _ _ _ _	do.
Short trees:			
Boxelder	25	Fast_ _ _ _ _ _	Short
Russian-olive	25	_ _ _do._ _ _ _ _	do.
Russian mulberry	20	Medium_ _ _ _	do.
Osage-orange	25	_ _ _do._ _ _ _ _	Long
Rocky Mtn. cedar	25	Slow_ _ _ _ _	do.
Shrubs:			
American plum	10	Fast_ _ _ _ _ _	Medium
Chokecherry	10	_ _ _do._ _ _ _ _	do.
Tamarisk (south only)	10	_ _ _do._ _ _ _ _	Short
Common lilac	8	Medium_ _ _ _	Medium
Caragana	8	_ _ _do._ _ _ _ _	do.
Skunkbush sumac	6	Slow_ _ _ _ _	do.

MEDIUM TO DEEP UPLAND SOILS (SANDY LOAMS AND LOAMY SANDS)

Mature size and common name	Maximum height (feet)	Height growth	Useful life
Tall trees:			
Cottonwood	60	Fast_ _ _ _ _ _ _	Short
Siberian elm	50	_ _ _do._ _ _ _ _	Medium
Shortleaf pine (south only)	50	Medium_ _ _ _	do.
Ponderosa pine	60	Slow_ _ _ _ _ _	Long
Austrian pine	50	_ _ _do._ _ _ _ _	do.
Medium trees:			
Green ash	30	Medium_ _ _ _	Medium
Bur oak	35	Slow_ _ _ _ _ _	Long
Eastern red cedar	35	_ _ _do._ _ _ _ _	do.
Short trees:			
Boxelder	20	Fast_ _ _ _ _ _ _	Short
Russian-olive	20	_ _ _do._ _ _ _ _	do.
Russian mulberry	15	Medium_ _ _ _	do.
Osage-orange	20	_ _ _do._ _ _ _ _	Long
Rocky Mtn. cedar	25	Slow_ _ _ _ _ _	do.
Shrubs:			
American plum	8	Fast_ _ _ _ _ _ _	Medium
Chokecherry	8	_ _ _do._ _ _ _ _	do.
Common lilac	8	Medium_ _ _ _	do.
Skunkbush sumac	5	Slow_ _ _ _ _ _	do.

FIGURE 5. *Possibilities of Shelterbelt Planting in the Plains Region.* The federal government published and disseminated a manual of operations, established by expert recommendations, including species, that tactically advanced the project of afforestation. Note the recommendation of the mislabeled Siberian elm (*Ulmus pumila*) across different soil types. Lake States Forest Experiment Station, U.S. Forest Service, *Possibilities of Shelterbelt Planting in the Plains Region* (Washington, D.C.: Government Printing Office, 1935).

The regulatory environment stimulated the vast expansion of mislabeled "Chinese elm" as effectual stock was conducted to federally owned nurseries in order to germinate plantlets. In other words, the human project was a success for *U. pumila* because they spread and germinated more seed. The movements of seed suggest samaras ride winds, float in streams, get stuck on the wheels of tractors, are picked up by birds, and nestle themselves on furry animals. In the context of the federal distribution networks, migration included a collaboration with expert foresters. The cultivation through local nurseries and the dissemination by the Civilian Conservation Corps (CCC) secured a lasting relationship between humans and *U. pumila,* between woody, shallow roots and the extant, underground grassland network.

The success of humans as a dispersal agent is a feature of *index.* Accurate recordkeeping cites the success of planting, as *U. pumila* triumphs over a planting zone of 17,000 miles.[28] In order to achieve such scales, seed was collected when it ripened in the spring or early summer and redistributed to centrally located seed stations in each state that would store, sort, and clean stock prior to spring sowing.[29] Such human–plant interrelations are scalable because the plantlet is calculated as a unit, without movement or interdependencies, easily gaining traction through statistics. But the seed source and its slippery progress evade metrics, as behavior falls between the cracks of territorial control, typology, and ownership lines. Effectiveness was measured by the quantity of seed and rate of growth, qualities that *U. pumila* confirms easily, lending its vigor and fecundity to the first organized, federally funded afforestation project.

This is precisely how foresters and plants remade the prairies. Measuring impact in miles and quantity corresponds to a loose category defined mainly in additive, linear terms such as seed bulk or planting area. As a result, the topic of soil erosion was never amended with the traits and behaviors of grasses and forbs. In more current terms, *U. pumila* costs the federal government $100 million a year in elimination costs that do not exactly remove the tree but control it by stump girdling, chemical injection, and setting small fires to kill saplings.[30] Because of invasion success, *U. pumila* is excluded from current species recommendations that define sanctioned and unsanctioned categories. Leaving the discontinuities out of past equations means that behavior can come charging back into the present—unaccounted for and unsolicited. For federal foresters, and developer farmers largely preoccupied with sedentary agriculture, *U. pumila* afforded a guided entry to the natural world. However, the life of *U. pumila* is temporally more colored by time, as plant life insists on adapting and extending its range for survival, an irresistible advancing tendency that rides the waves of human technology. The behavior of *U. pumila* works well with scientific inquiry; as standardized project elements are established, it conforms and settles in for the long haul.

SUPERWEEDS

Plants that disperse seeds rapidly and profusely tend to grow particularly well in exposed micro-environments, eagerly seizing barren, unprotected ground. These plants are often called weeds. *Ulmus pumila* is considered "invasive," another word for weed, a term that resists single definitions. Invasive-plant literature includes "exotic," "nonnative," "spreading," "pest," "non-indigenous," "incursive," and "alien" with inconsistent meanings. Yet the term *invasive* seems to spread most rapidly, where it thrives as a niche subject of the conservation movement. Consider the success of *Ulmus pumila* seed through the specialization of federal frameworks. According to the Plant Conservation Alliance (PCA), a partnership with the Bureau of Land Management (BLM), "for long-term management of Siberian elm, reduction of seed source is essential."[31] The mitigation goals enlarge along with the plant's ability to reproduce, as the PCA goes on to cite that *Ulmus pumila* is present in forty-one U.S. states, excluding Hawaii, Alaska, and a handful of coastal ranges, including New Jersey and Delaware. Is this invasion success?

Invasive plants are defined by the International Union for Conservation of Nature as "an alien species which becomes established in natural or semi-natural ecosystems or habitat, is an agent of change, and threatens native biological diversity."[32] Another definition of invasion could be a plant that is efficient at migration and lends itself to amassing communication signals between germination and dormancy. Ultimately, either description defines behavior, but the terms *natural* and *native* do little to include the persistence of human disturbance. Rather, these terms are culturally produced, not biologically specific.[33] A more inclusive definition would acknowledge plant life by rendering specificity in behavioral terms specific to manifold species, such as rate of growth, range of dispersal, and communication mechanisms.

Botanical historians agree that the first "weeds" emerged with the advent of agriculture; before that, these curious plants were "secondary succession colonizers" in ecological terms. Such plants specialize in reoccupying the ground once stripped of plants by landslides, floods, and fires, so they were poised to do their work when tilling and plowing became the next disturbance. Yet the frustrations between "invasive" and "native" plants percolate into the channels of practical use, reiterating international legal frameworks and expending billions of dollars to eradicate weeds. These same agencies also delineate which species are threatened by "alien invasives," as subsets of threat fall into three categories: vulnerable, endangered, and critically endangered.[34] Given the contemporary currency of establishing what is and is not "nature," it is worth elaborating on the costly conundrum that positions invasive plants as a global debate of winners and losers.

In the United States, estimates assert that species introductions exceed 50,000 individual alien invasives.[35] Of these, the more useful or productive plants are often called "introduced," species that include corn, wheat, rye, and other food crops. Cultural dependence on plant life determines the dialogue, subjecting select plants to categories of productiveness to the human project. Yet it is so much more than dependence at stake in activating a more nuanced approach to plant life because accepting behavior exposes the simplistic currency of the past: plants that take well to domestication are A-listed and plants that do not take well to domestication are taken off the list, or put on other ones. For instance, plants "native" to the United States or recorded through the lens of Europeanization tend to fall on the invasive list, clearly outlining the inconsistencies and past abuses. Consider the herbaceous plant *Amaranth palmeri,* a foundational prehistoric crop and a staple of Aztec religions.[36] *Amaranth* produces 250,000 to 500,000 seeds per plant and mostly "invades" the American Midwest, even establishing itself as far north as Minnesota and as far south as Tennessee, engendering alarming statements such as "This weed is like no other, but tools exist to defeat it" or "A new weed threat to watch out for."[37] Coverage from a *New York Times* article calls it a "superweed" because it has evolved a resistance to the chemical glyphosate, commonly known as Roundup.

> Pigweed can grow three inches a day and reach seven feet or more, choking out crops; it is so sturdy that it can damage harvesting equipment. In an attempt to kill the pest before it becomes that big, Mr. Anderson and his neighbors are plowing their fields and mixing herbicides into the soil.[38]

These accounts bring to light the variety of claims over progress, science, and governance that exclude the human as an agent of dispersal and evolution. To study engagement requires turning away from the abstractions of purely human expertise, to tune into the adaptive tactics of plant life. Following the spread, emergence and cultivation of plants render a world variegated by the advance of sedentary civilization, which also means facing the consequences. The story of *Ulmus pumila* contours the history of settled agriculture—the turning of soils, the furrows of the plow, and the destruction of the grassland in favor of short grass crops, like wheat.

Unlike *amaranth,* many "invasive" or infamous plants were deliberately introduced to the American continent, including *Pueraria lobate* (kudzu) to stabilize soil in the south, or *Fallopia japonica* (Japanese knotweed) through the nursery trade and *Lythrum salicaria* (purple loosestrife) for its aesthetic value. *Ulmus pumila* is one of many plants introduced as a "solution" to the nutrient-poor soils of the agricultural prairie, and as a leading tree in the Prairie States Forestry Project. The act of collecting and germinating seed to establish federal nurseries, as a means to convey and transplant seedlings, was an operative tool of integration along the drylands of the 100th meridian. As we have seen, seeds sensed a similar biome in the aerial patterns of drought and rainfall, recognized the familiar sandy substrate, and detected the windblown, arid surfaces. While grasses retreat into their roots and trees thicken their bark in reaction to aridity, all plants migrate to find less-extreme conditions. The seeds of *U. pumila* were already completely adapted to the novel aridity brought on by agriculture, unlike the extant plants that were acclimatized to the moisture-rich rhizosphere of the prairie grassland. *Ulmus pumila* migrated successfully when it landed in the prairie states. Alone in a monoculture of annual seeds, it had one idea and one main task to accomplish: continue to extend habitat, push beyond the given range, and gain new territory.

Living Units

Most "invasive" plant species advance through a process of signaling called allelopathy. Allelopathy is a form of chemical communication that enables contact between plants, or discourages contact, creating a unique territory for particular relationships. As a result, allelopathy tends to affect the structure of extant communities by lowering diversity and abundance. Theophrastus mentions the inhibitory—or allelopathic—effect that some plants have on others, but only briefly.[39] Despite the imperceptibility of allelopathy, the inhibitions of plant life are as observable as our own retreat from poisonous products or dangerous organisms. Other species recoil from the signal. The word *allelopathy,* meaning "mutual harm," is expressed as compounds, discharged directly into the soil by roots or through multiple decomposing residues left on the surface, such as fallen bark or seasonal leaves.[40] This residue is often termed "litter," implying that such remnants are a waste product, separating the plant from the animated soil. But as we have seen, an individual plant or tree unit cannot be dislocated from context without repercussions. Allelopathic transmissions are one adaptation that a plant might acquire slowly in another extreme circumstance and then deploy quickly once introduced into a novel environment. It is an adaptation that confirms plant life beyond unit calculations and attests to aliveness. Plant life communicates outside individual growing points, extending ecologically through prolific seed, leaf litter, migrating fruit, and chemical boundaries.

Subtle gradations appear and persist around the growing point, appearing barren or plant free, thus augmenting the potential for erosive forces across the upper horizons of the soil. In other words, chemical survival mechanisms expose land, which lends itself to further desiccation that eventually becomes more difficult to cultivate. Imbued with the power to protect themselves, invasive, allelopathic plants are often the bane of agricultural systems. The relations of plant–plant or plant–predator and plant–human exchange are explicitly located in this type of elusive relationship, an exchange primarily located in the rhizosphere, as plant–microorganism and plant–mycorrhizae multiply each relation. When allelopathic communication occurs in the leaves and in the atmosphere, the majority of the biochemicals that influence growth land on soil surfaces and slowly seep underground.

In the story of *Ulmus pumila,* the release of allelo compounds from so-called litter is registered in the rhizosphere, assimilating and exerting influence over microorganisms and inhibiting the establishment of other, perhaps more local, tree species, or cultivated crops and annuals.[41] Chemicals mix and mingle in the rhizosphere, which is further influenced by root exudates, the sloughing off of root tissues and cells. The rhizosphere is an area of high microbial activity, lower pH, increased organic matter, higher carbon dioxide, and lower water and nutrient content.[42] Appreciating the extents of plant life beyond the unit requires a shift to a vertical imagination, one that connects the atmosphere and the rhizosphere. In much the same way, a change of perspective that appreciates rhizospheric formation helps resist the planning, divisions, and categories that drive a line through the study of plant life and disregard chemical communication.

Living soils are composed of a cosmopolitan array of microorganisms, such as microfauna, fungi, insect larvae, and mycorrhizae. The estimated number of soil fauna in an average, mineral soil profile is 1.5 million/m^2, yet ranges of up to 120 million/m^2 have been attributed to remnant grassland soils.[43] The buried flows of soil micro-fauna expedite decomposition, aerate the soil, and feed from roots and rhizomes. The swarm of plant interactions is woven into a dense network of chemicals, pathogens, minerals, clays, and moisture, adhering, clustering, and repelling influence. Within this active "society," the soil atmosphere also contains amounts of nitrogen, carbon dioxide, oxygen, and water vapor.[44] Oxygen is in high demand for roots, while roots are exerting significant influence on the micro-fauna and vice versa. Attention to the rhizosphere confirms that plant life defines the scope of interaction, whereby roots, rootlets, rhizomes, tubers, bulbs, and seeds are affecting stringent boundaries by inviting collaboration.

Across the prairies, grasses deliver the bulk of the soil profile, as a square yard of prairie holds twenty-five miles of rootlets.[45] But the diversity in this deep profile is increased by the forbs—the leafy plant—which fix nitrogen and minerals, creating fertility in the soil. Cooperation and coproduction are the motor of the prairie, corresponding to an average of thirty-year drought cycles, regular to the grasses and forbs but catastrophic to industrial farming. During drought cycles, the border between tallgrass prairie and shortgrass prairie shifts hundreds of miles in slow motion, like a tide. Some plants disappear altogether from the surface and live on in root systems, which lie dormant in the dry ground. Retreat and advance are common; seed rises to the wind and the air and finds less-extreme peripheries. This motion in time and space is the survival of the prairie organism. Trees that cannot account for the drought or exploit the wind are clearly distinguished from the undulatory, continuous, and cyclical prairie. In "The Living Network in Prairie Soils," Weaver lays out this long, sinuous formation as a predictable but unstable boundary.

> Soil is the unconsolidated outer layer of the earth's crust which, through the processes of weathering and the incorporation of organic matter, becomes adapted to the growth of plants. It is usually underlain by unconsolidated parent materials into which the deeper roots of plants frequently extend. The soil often contains and acts upon a much more extensive portion of the plant body than does the atmosphere. This is especially true of grassland soils.[46]

Grasses grow from buds close to or even slightly buried beneath the ground and can recover from disturbance such as fire or grazing with ease. They also have long leaves that grow by pushing out from their base, which means that they can regrow very easily if the tops of the leaves are burnt or eaten. In fact, most grasses seem to "need" frequent fires or grazing to keep other plants out.[47] In the absence of either type of disturbance, grasses are eventually outcompeted by woody plants such as trees or shrubs. *U. pumila* acquired the skills to endure through disturbance from its extreme sandy-land circumstances, enabling prompt settlement across the American prairie. Unlike earlier, "European" imports, these Eurasian grassland species were completely comfortable with the shifting climate and the disruptions of aridity. *U. pumila* colonized effectively, reorganizing the landscape in conjunction with the soil. In so doing, *U. pumila* engenders the vast depth relations of the rhizosphere that annual crops do not locate, making and remaking specialist inquiry. Human migration and settlement remain a powerful and active dispersal agent for plant life, evidenced in the speed of *U. pumila* seed. Plant life emerges from pure commercial instrumentation into the wonders of compromise, a feature of collaboration.

TRACE

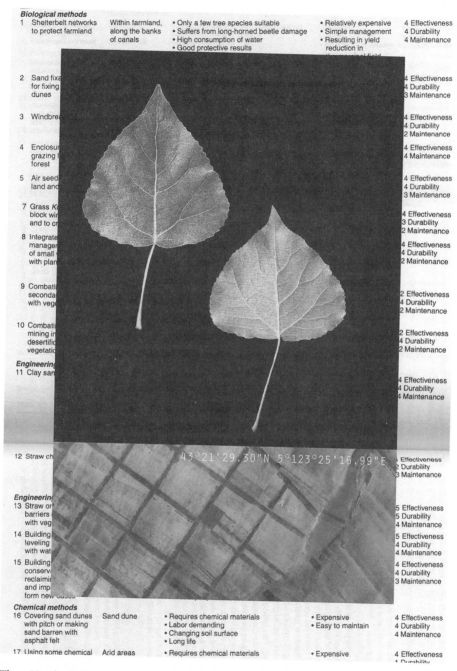

Biological methods

1 Shelterbelt networks to protect farmland — Within farmland, along the banks of canals — • Only a few tree species suitable • Suffers from long-horned beetle damage • High consumption of water • Good protective results — • Relatively expensive • Simple management • Resulting in yield reduction in — 4 Effectiveness / 4 Durability / 4 Maintenance

2 Sand fixa... for fixing dunes — 4 Effectiveness / 4 Durability / 3 Maintenance

3 Windbre... — 4 Effectiveness / 4 Durability / 2 Maintenance

4 Enclosu... grazing... forest — 4 Effectiveness / 4 Maintenance

5 Air seed... land and — 4 Effectiveness / 4 Durability / 3 Maintenance

7 Grass *K...* block wi... and to cr... — 4 Effectiveness / 3 Durability / 2 Maintenance

8 Integrate... manager... of small... with plan... — 4 Effectiveness / 4 Durability / 2 Maintenance

9 Combati... seconda... with veg... — 2 Effectiveness / 4 Durability / 2 Maintenance

10 Combati... mining i... desertifi... vegetatio... — 2 Effectiveness / 4 Durability / 2 Maintenance

Engineering...
11 Clay san... — 4 Effectiveness / 4 Durability / 4 Maintenance

12 Straw ch... — Effectiveness / 2 Durability / 3 Maintenance

Engineering...
13 Straw or... barriers... with veg... — 5 Effectiveness / 5 Durability / 4 Maintenance

14 Building... leveling... with wat... — 5 Effectiveness / 4 Durability / 4 Maintenance

15 Building... conserv... reclaimi... and imp... form new oases — 4 Effectiveness / 4 Durability / 3 Maintenance

Chemical methods
16 Covering sand dunes with pitch or making sand barren with asphalt felt — Sand dune — • Requires chemical materials • Labor demanding • Changing soil surface • Long life — • Expensive • Easy to maintain — 4 Effectiveness / 4 Durability / 4 Maintenance

17 Using some chemical — Arid areas — • Requires chemical materials — • Expensive — 4 Effectiveness / 4 Durability

43°21'29.30"N 5°123°25'16.99"E

Three Norths Shelter System imagined by the procedures of artifact, index, and trace. Image by the author.

Trace accelerates afforestation in material terms. The connotation of the word insinuates an outline or a delineation by which lines are figured, lines that pull *artifact* and *index* into the world, into a sociospatial milieu. *Trace* is distinguished on the ground by an infrastructure of planting that marks, scars, and inscribes afforestation. Success and failure lie in the remnants of dryland development, the register of nurseries, rows, grids, belts, and blocks. *Trace* is not only something we can see; it is also a way of seeing planting when it is rationalized to gain global credit. Consequently, *trace* refers to the evidence of tree planting radically disconnected from context, the landscape of afforestation.

The landscape of afforestation is a synthetic construction, entirely made up by the surveys, maps, and objectives of reliable records, experiments, and data sets. Here, landscape relies heavily on units sanctioned by enthusiastic specialists. Landscape without afforestation is defined as the proximate ground and air that support terrestrial life on our planet. Landscape is rendered through plant life, experienced as a bond between the atmospheric climate and the density of underground relations. The qualities of the landscape are temporal and earthly, the messy context absent on a crowded tally of number units and sanctioned species lists. In a practical sense, landscape is lived.

Under the disguise of environmentalism and conservation, *trace* sanctions tree-unit planting under the auspices of aid and relief. It projects stability even though economic, social, and ecological circumstances change. Trees either persist or die as phenomenon mingle and progress, while cultures mature and climates change. As failures are reported, *trace* expands to find proof, activating ever-increasing networks of nurseries, research stations, and plantation experiments. Each breakdown offers the opportunity for improvement, as mandates are amended to keep up with current crises. Once vindicated, projects multiply without question because it is relatively easy to plant millions of trees. By comparison, it is remarkably difficult for millions of trees to grow, thrive, and survive in drylands.

Notable to afforestation as a globally sanctioned "aid" procedure is how tree planting endorses fixed settlement in largely nomadic cultures, intruding on routines that evolved with treelessness and seasonal aridity. Diana K. Davis, scholar of dryland environmental history, persuasively states:

> China is one of the many countries hoping to increase food production by enforcing agricultural expansion into its arid and semiarid zones, most of which are marginal for agriculture and were traditionally regions of extensive nomadic pastoralism with long and successful histories.[1]

Davis indicates that the failure of dryland development emerges from the notion that aridity is necessarily negative and that the desert is predominantly uninhabitable. I am suggesting that development projects fail because of the global indifference toward plant life, notable because planting is the most significant contribution to anti-aridity projects. Davis rightfully proposes that schemes to augment forestation or agriculture fail due to the perception of the desert. I suggest that afforestation replaces landscapes of lived experience, practical knowledge, and sensible care of the plants that are actually adapted to drylands, which leads to failures in forestation and agriculture. Collectively, the negative image of the desert and the failure to notice plant life circumscribe settler societies, the advance of European invasion, colonial rule, state control, or carbon-copy conservation. Political authority inflects ecological concepts to speed up and enable control by excluding plant life, a tradition that continues to inform environmental policy and is sneaking its way into climate policy.

Technology is a feature of *trace* discernible throughout the Three Norths Shelter System (TNSS), following fallow fields sedimented by industrial agricultural, abandoned irrigation canals, redundant drip-line infrastructure, and provincial sink holes that drop far into the sandy substrate. The control of technology is regulated by the state, as standardized typology advances the project by suppressing aeolian transport, biotic movement, abiotic rhythms, and the social patterns that swarm around each design. This afforestation project is unique because it spans decades, and each decade is managed in phases, while each phase is bracketed by improved planting techniques. The TNSS is a composite of issues, productions, and contexts that endures today.

Populus spp. is a tree technology. It fosters research and incites planting because it is amenable to unit isolation. It empowers afforestation and provides evidence, as the TNSS proliferates across 42 percent of the country. At the same time, *Populus* spp. unsettles phases by multiplying atmospherically through airborne pollen and biologically as underground rhizomes that produce clones. The politics of plant life echoes in the remnants of each phase, although material evidence is extraneous to plant life. The matter of concern is that the compulsion toward billion-tree programs is no different today than it was during the turn-of-the-century forestry. What we have now is simple exploitative twenty-first-century afforestation.

7

Contextual Indifference

The essence of the problems of raw materials and development is the struggle of the developing countries to defend their state sovereignty, develop their national economy, and combat imperialist, and particularly superpower, plunder and control. This is a very important aspect of the current struggle of the Third World countries and people against colonialism, imperialism, and hegemonism.[1]

—Deng Xiaoping, 1974

IN 1974 DENG XIAOPING addressed the United Nations General Assembly with a sequence of claims that would set the stage for China's emergence as a superpower. The vision of Three Worlds articulated a new approach to world affairs: The United States and the Soviet Union belonged to the First World. Countries such as Japan and Europe were part of the Second World. All "underdeveloped" countries constituted the Third World, including China.[2] In putting forward the national interest of China, Deng elaborated the needs particular to China and the continent of Africa, accusing privileged countries (the United States, Russia) of excluding developing nations in the world order. The relationship of Deng's globally focused argument to the local relevance of *raw materials* was a brilliant way to rationalize further national authority over the expansive landscapes of western China. The speech strategically aligned China with other Third World nations and caught the attention of the United Nations.[3] It also inspired the Chinese public to feel connected to global politics, immersing Deng's future governing strategy in national economic development.

Over the next decades, Deng engineered China's political, economic, and social life, most memorably by establishing the one-child policy, setting up special economic zones, and endorsing the market economy. Each reform notoriously operated across the continent, with widespread force and the driving ambition credited to Deng's character. His alliances with the United Nations enabled a growing market and correspondingly included China's emerging environmental sector. In particular, China's Four Modernizations policy promised to develop the fields of agriculture, industry, science, and defense.[4] With global alignments in mind, Deng effectively maneuvered environmental policy and economic development. One relevant modification shifted the focus of the Ministry of Forestry (State Forestry Administration, SFA) from its agricultural origins in fiber production to an agenda dedicated to securing timber resources across the nation. The modification was even more specific, as timber resources were intriguingly defined by securing forest cover.[5] In particular, Deng devised a signature project that could compete with the ambitions of other reforms and help modernize agriculture: a protective forest system in the Three Norths area.

At the time, it was unclear what the region was being protected *from.* Nonetheless, Deng founded what is currently called the Three Norths Shelter System (TNSS), solidifying the first signs of supracontinental afforestation: a national policy with incalculable repercussions that rested on the vast temporal revelation of proposing a timeline that far exceeded his power. The TNSS is planned along a seventy-year timeline, split into three phases, and maintains the ambition to afforest more than four million square kilometers, in 551 counties within 13 provinces.[6] The area accounts for 42 percent of Chinese land: a conception of tree planting appropriated from the United Nations first map of desertification, a manipulation of context that assumes all dryland is unoccupied and free for planting.

This chapter focuses on the effects of afforestation on the landscape. Because it is the final section, it is slightly different from the rest of the book in that it looks to the future of afforestation using the immediacy of the TNSS. In part, this chapter suggests that current acts of "saving" continue along the trajectory outlined in the other cases, indicating that the negotiations of the present implicate policy in the contested past. Despite its awards and accolades, the TNSS can be placed in the "solution" tradition because it alleviates the pressure of conservation without the attendant risks or rebellion of actual change. Nothing is lost, destroyed, or interrogated. The tensions are depleted by tree planting. But there is more to this, and in order to understand it more clearly, the backdrop must be examined, circulating back to the global mandates authorized by the creation in 1972 of the United Nations Environment Program (UNEP), which stimulated international interest in drylands at the height of the Sahelian drought (1968–72).[7] The magnitude of the drought precipitated years of famine, widely characterized as the first global crisis before ozone depletion or climate change.[8] The wide-ranging effects of environmental decline were global by definition, precipitating extensive political cooperation.

Given the alliance between afforestation and conservation, it is not surprising that the response to the Sahelian drought was caught up in tree planting. Sanctioned action is epitomized by planting trees, the proof mediated by *trace* to engender afforestation. The immediate decades following the establishment of the UNEP represent a marked changed in the language and laws governing the environment, especially in relation to drylands. According to Diana K. Davis, "The ideas of desert and desertification embodied by the UNCCD, most national governments, and innumerable NGOs are fundamentally flawed and have triggered many misunderstandings that have led to numerous misguided policies and project failures."[9] Despite increased criticism and systemic misunderstandings, tree planting in drylands endures as standard policy, as global problems require global solutions. At the time, the strategy of conservation is established by rearranging the inventory of management techniques, including afforestation. To reiterate the suggestion from the introductory remarks, because it is unclear who gains, tree-planting programs keep humans safely on the outside of any crisis, charting an achievement that is only integral to some portion of society. As a result, when a simplified version of tree planting is arrayed across vast dryland territories, it reveals more about human relationships than it does about environmental ones.

In this case, human relationships assembled around the United Nations *Desertification Map of the World*, the first record to suggest the global extent of desertification (see Figure 6). This map exemplifies the universalization of environmental crisis, unifying the claim that deserts are encroaching on "productive" land across the world. There are multiple ways of framing the conflicts that arise on this map. First, it reveals the tension between the spatiality of geography and the aspatiality of botany. This is achieved by mapping techniques that label or categorize treeless or so-called barren lands. Second, while it claims to be a map of desertification, it could also be a map of drylands. Since vegetation maps reduce complex landscapes to the average density of leaf foliage—also referred to as *cover*—forest types dominate the representation, as tree canopy is translated into a consistent hatching pattern whose edges are defined by clear, dashed strokes. The result is that thousands of plant species are indistinguishable from one another, obviously disregarding the roots and rhizomes living under the ground, the vines that creep vertically, and the annual cycles of budbreak and seed dormancy hinged on diverse climates. Another way to frame the conflict is the undertaking of recording nature that establishes a distinct authority, as order is constructed from apparent disorder, translating complexity into convenient and manageable units.[10] The map simplifies at the landscape scale, suggesting a hierarchy of value; dryland regions (prairies, grasslands, savannas) have less value because they offer less coverage, and thus such hackneyed concepts as the "naked" desert arise. Thick, dense hatching celebrates the verdant forest, and scattered points scorn the barren desert, construed as a measure of deficiency. On a vegetation map of the planet, there is no life where there is no hatch.

The tree-planting operations proposed by the TNSS were originally sanctioned with UNEP as the authoritative collective. For instance, in 1977 the first call to unite the global "problem" of drylands was legitimized by the UN Conference on Desertification, a meeting instigated by growing anxieties following the great Sahelian famine.[11] The conference was distinguished by

FIGURE 6. The first map of desertification unified drylands across the planet, despite radically different biotic and cultural contexts. Compiled by the United Nations Environment Program (UNEP). Food Agriculture Organization of the United Nations, UNESCO, and World Meteorological Organization, *Desertification Map of the World / Carte mondiale de la désertification* (Nairobi: UNEP, 1977).

the innumerable national development agencies that came together to review the status of desertification at a scale of one to a million.[12] At this scale, the American prairies, the sub-Sahel, and the western fringes of China shared a common "hatch" that demarcated levels of concern. The UN map of desertification was the first of its kind. Appropriately, the launch date of the first phase of the TNSS coincided with the launch of the map in 1977, propelling China into global affairs.

While Deng aligns the project with desertification and "hyper" aridity, the project is ongoing to this date, and skillfully shifts its mandate with each passing global paradigm, satisfying the UN's desire to imagine China doing its environmental part. Since then, China's shelterbelt project is skillfully vindicated through each decade of environmental decline, including soil erosion, water security, and air pollution, in addition to desertification. In each case, an organized effort to plant trees in otherwise treeless biomes is offered as a conservation measure, a solution. More prominent responses to environmental decline such as halting unsanctioned deforestation, or limiting industrial contaminants, massive overgrazing, or urbanization, are not as visible or as popular as tree planting. To reiterate the introductory remarks, afforestation is produced by a trajectory of understanding that pairs environmental action with environmental destruction. Tough environmental choices are mostly sidestepped when terms like *fix* or *solve* emerge on the global stage.

Afforestation produces evidence for policy makers. It is physical and material proof that global-aid networks are effective. It engages specialization and grounds science. Tree planting is reduced to a replicable procedure, pacifying aliveness to shape a construction detail designed to proliferate across vastly differentiated landscapes. For instance, the effort of planting typically exhausts the social frontiers of the local. It relies on a global agreement ratified to plant trees by activating a regional network of nurseries, suppliers, irrigation consultants, and engineers. Trees are substantiated in the "real world" through standard planting techniques, select propagation trials, and distribution networks. While units negate the specificity of the plant, standardization through typology negates the specificity of the soil. Plants are selected for their ability to reproduce and replicate, producing an endless supply of convenient units. The biotic ground is reduced to a calculation of resources, amassed as stock and deployed in cubic meters.

It is not only the biotic profile that is disregarded by typology; deeply cultural meanings attached to land likewise erode through conservation-based development. The rejection of context is sanctioned by global aid and, unlike the American shelterbelt project, refined by international organizations and their member states, revealing the same contradictions that continue to pervade global environmental debates. Concerns are dutifully abstracted from reality and absorbed by economic development.

China is a major force in Africa, rivaling the United States in some countries. In 2009, China became Africa's largest trading partner, passing the United States. Sino-African trade reached $128.5 billion in 2010. Chinese investment in Africa, although still modest compared with European investment, is growing faster than Western investment. Total aid to Africa is still small compared to that of Western countries—averaging perhaps $2 billion annually in the last few years. By comparison, assistance to Africa from Development Assistance Committee countries is running at about $30 billion annually. On the other hand, China is offering huge low-interest loans, often tied to infrastructure projects built by Chinese companies and paid for in natural resources shipped to China. In some countries, these loans surpass the total of all loans from other countries or international banks.[13]

According to diplomat and historian David Hamilton Shinn, trade was the first link between China and Africa. Since then, China has invested in a model of "prosperity and stability" to safeguard its interests in Africa, chiefly through the Forum on China–Africa Cooperation (FOCAC), established in 2000. The Chinese approach is "packaged" with the assertion that Africa can provide for China's growing needs, and in return offers monetary and political support. This "peaceful development," or "rise," as FOCAC terms it, aims to encourage mutual confidence and internal cooperation between China and its African partners as Chinese workers, business owners, and officials move to Africa. The growing interest in Chinese foreign affairs highlights concern for *development* by displaying a willingness to engage in international activities such as UN peacekeeping operations.[14] China's strategy toward Africa includes, among other things, increased trade between Africa and China, through technology, education, and cultural cooperation; and to work closely with African counties in multilateral organizations such as the UN and the WTO, in order to address climate change, food security, poverty alleviation, and rural improvements. The financial support and expertise provided by internationally endorsed development use two familiar conservation tactics: agroforestry and afforestation. Both models enhance the contingency of the encounter between trade partners through plant life. Notable are the specialists: institutional organizations, foundations, and nongovernmental organizations that trade in units in order to inscribe policy across the landscape.

Development is often coded by cultivation. It can only be decoded by the "productivity" narratives proffered by international organizations, government departments, charities, and aid movements. Land that is not productive can become productive through investment and technological expertise. In these simplified relations, development is cultivation, and cultivation is agriculture. Thus, a loan-burdened infrastructure project can be paid for in natural resources shipped to China. In many cases, the tendency toward productivity creates homogenous holdings out of diverse landscapes. Policy tends to be followed by claims that there is no productive land or that land can be made productive, tracked by development incentives. The question of productivity is not an ecological one—it is an expansionist, colonial, and capital model that advances through contextual indifference, and continues to distance human and plant life.

LINES OF POLICY

The Aihui-Tengchong line is a vector traced across a continuous forty-five-degree angle from tropical southwestern China in Tengchong, Yunnan, to the very north of Chinese territory in Heilongjiang province, Heihe. The demarcation, colloquially referred to as the Hu Line, divides the population of the country, drawing a striking difference in the distribution of China's population.[15] It is radically indifferent to topographic specificity but reveals the significant affiliation

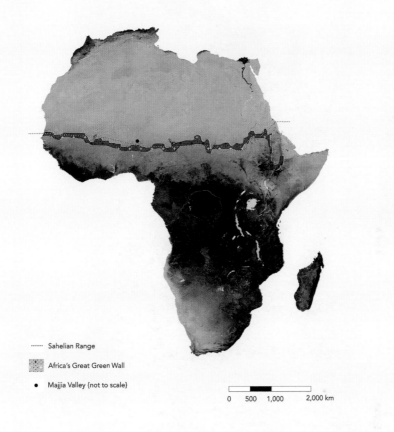

Sahelian Range

Africa's Great Green Wall

Majjia Valley {not to scale}

0   500   1,000   2,000 km

PLATE 1. Africa's Great Green Wall, 2015–present. The African project advances afforestation as a supranational, cross-continental initiative to unify eleven countries along the sub-Sahelian range. Map by the author.

PLATE 2. *Faidherbia albida.* A herbaria specimen archives taxonomic confusion. In this image digitized in 2014, *Faidherbia albida* was originally labeled *Acacia Senegal* and corrected to *A. albida. Faidherbia* is also known as "balazan" in Mali and "saas" in Senegal. Image courtesy of the Natural History Museum, London.

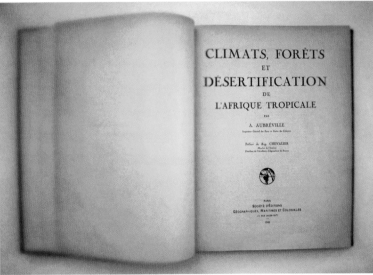

CLIMATS, FORÊTS
ET
DÉSERTIFICATION
DE
L'AFRIQUE TROPICALE

PAR

A. AUBRÉVILLE
Inspecteur Général des Eaux et Forêts des Colonies

Préface de Aug. CHEVALIER
Membre de l'Institut
Président de l'Académie d'Agriculture de France

PARIS
SOCIÉTÉ D'ÉDITIONS
GÉOGRAPHIQUES, MARITIMES ET COLONIALES
17, RUE JACOB (VIᵉ)
1949

PLATE 3. *Above*: There is little to show for the Great Green Wall of Africa apart from small pilot projects that draw attention to the act of planting individual trees. In this image, Senegal's environment minister celebrates afforestation surrounded by Senegalese and French military personnel in the village of Labgar, November 2009. Photograph by Universal Images Group via Getty Images. *Below*: André Aubréville was a botanist commissioned to catalog plants for L'Afrique Occidentale Française (AOF) in colonial West Africa. Aubréville published the term *désertification* for the first time in *Climats, forêts et désertification de l'Afrique tropicale* (1949), asserting that the condition was produced by exploitative forestry practices in subtropical biomes, not natural processes across drylands. Image by the author.

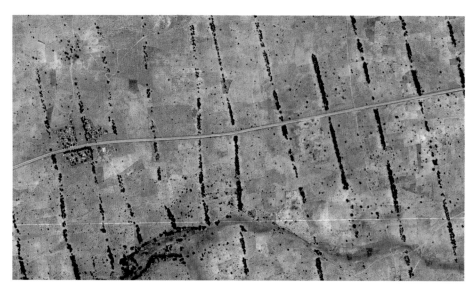

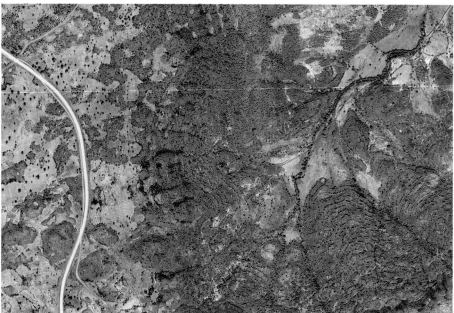

PLATE 4. *Above*: Approaching Majjia Valley, southern Niger. Afforestation projects are spatially aligned with road and rail infrastructure. The linear shelterbelt displays a hyperlocalized effect, though the typology contributes to total numbers when accounting for the project as a supracontinental effort. Image from Google Earth, 2010. *Below*: Majjia Valley, southern Niger. The Majjia windbreak project acknowledges the extant landscape, as the planting locations are aligned with the topography rather than infrastructure. Plant selection recognizes species that are familiar to farmers and villagers. Tree planting in drylands thrives with the care found in stewardship. Image from Google Earth, 2019.

PLATE 5. *Above*: A tree nursery in Niger grows plants as products to account for individual units, while spontaneous and mostly leafless *Faidherbia* trees thrive in the background of the plantation. Afforestation relies on growing trees that require artificial support, visible in the drip-line irrigation infrastructure. Photograph by Universal Images Group via Getty Images, 2007. *Below*: *Zaï* pits are micro-basins formed by soil and stone bunds that prepare the ground for plant life. The careful assembly considers generous spacing and is crafted by hand, requiring no additional mechanization or irrigation infrastructure. Photograph by USAID in Africa, April 2016, public domain, via Wikimedia Commons.

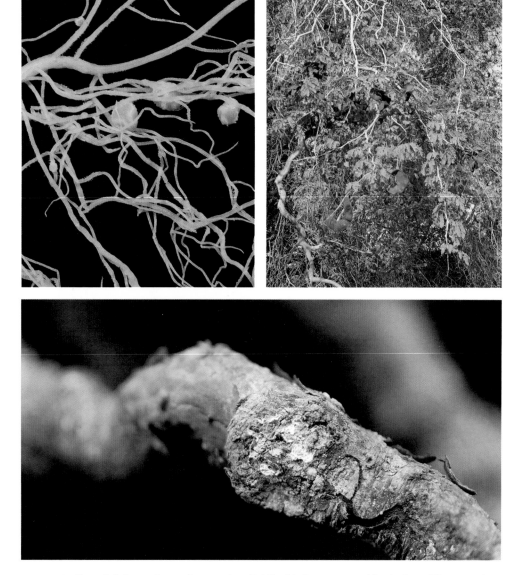

PLATE 6. *Above, left*: Leguminous plants such as *Faidherbia* form partnerships in the soil between their roots and soil dwelling *Rhizobium,* which appear within rootlets as round, durable nodules. Photograph by Praca własna, licensed under the Creative Commons. *Above, right*: *Faidherbia* belongs to a large family of leguminous plants that carry edible seeds in pods. Legumes are excellent at fixing nitrogen due to a symbiotic relationship with soil-dwelling bacteria. Plant and bacteria life exchange nitrogen for carbohydrates. Photograph by Ji-Elle, licensed under the Creative Commons. *Below*: *Mycorrizal fungi* live in the white, powdery undercoat of plant life, concealed in the rhizosphere. These organisms are barely visible in the soil but encourage chemical communication between plants. Photograph by the author, 2020.

PLATE 7. *Faidherbia* sheds its leaves at the beginning of the rainy season, August 2013. Light and moisture penetrate the soil and support seed development without overshading crops as they mature. Raquel Maria Carbonell Pagola, "Giant acacia trees in the savannah . . . ," courtesy of Getty Images.

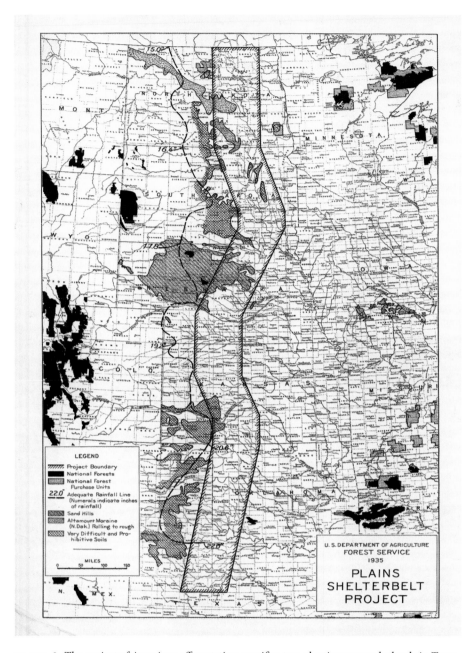

PLATE 8. The project of American afforestation specifies tree planting across drylands in Texas, Oklahoma, Kansas, Nebraska, South Dakota, and North Dakota, also called the prairie states. The project boundary is indicated by the hatched line and is variously called the Plains Shelterbelt Project, the Great Plains Shelterbelt Project, and the Prairie States Forestry Project. Map from Lake States Forest Experiment Station, U.S. Forest Service, *Possibilities of Shelterbelt Planting in the Plains Region* (Washington, D.C.: Government Printing Office, 1935).

PLATE 9. The Prairie States Shelterbelt Project, 1934–42. The American project was a New Deal program that positioned afforestation as an environmental demand, although it was a social policy. Map by the author.

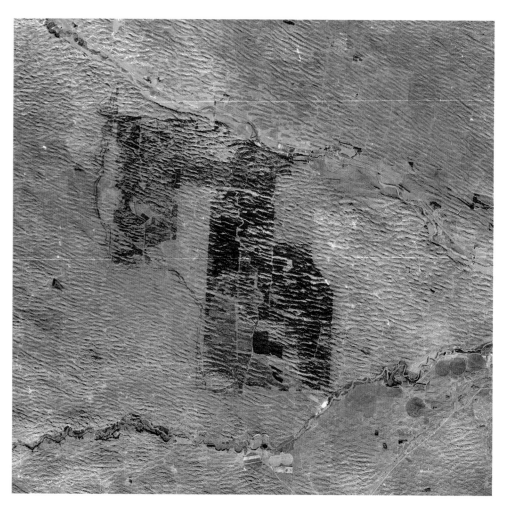

PLATE 10. The evidence of afforestation materializes as the first artificial national forest, planted by hand across the Nebraska Sandhills. The Bessey Nursery site conveniently establishes a stable water supply from the Middle Loup River. Irrigation is a requisite to successful experimental trials, pilots, and the triumph of afforestation typology. Image from Google Earth, 2016.

PLATE 11. *Above*: In *Sixteen Maps Accompanying Report on Forest Trees of North America* (Washington, D.C.: U.S. Census Office, 1884), Charles Sprague Sargent created maps of the United States showing the relative or average density of existing forest cover. Note the absence of tree cover along the 100th meridian. Map courtesy of the U.S. Department of the Interior, Census Office. *Below*: The Sandhills of Nebraska are largely treeless, a condition that is typically met with tree-planting efforts requiring that crews dig individual holes into prairie strata. Image courtesy of USDA archives, 1908.

1/27/1928 Three year old Chinese Elm – Plumfla's Nursery

PLATE 12. Afforestation remade the prairies through plant selection. At the time, Chinese elm (*Ulmus paviflora*) was confused with Siberian elm (*Ulmus pumila*) because it was collected in China. *Ulmus pumila,* often inaccurately called Chinese elm, still grows aggressively across the prairie states. Photograph courtesy of USDA archives, 1928.

PLATE 13. *Above*: The Bessey Nursery, shown here in 1907, was the first federal tree nursery, located in the otherwise treeless Sandhills. The nursery was a significant supply for the Prairie States Shelterbelt Project because it demonstrated decades of trial-and-error attempts at afforestation. Photograph courtesy of USDA archives. *Below*: Seedling production is a necessary first step of every afforestation campaign. Nursery infrastructure develops since procurement, purchasing, and ordering plants in the billions is not cost effective. Photograph courtesy of USDA archives.

PLATE 14. The label on the verso of this photo's mount reads: "A twig of the common north-Chinese elm densely covered with fruits. April 30, 1915" and continues: *Ulmus pumila L.*; *Ulmaceae*; Siberian elm; Chinese name 'Dya yu shu' meaning 'family elm tree.'" The description is evidence of the prolific seed habit of *Ulmus pumila* and the species confusion created by plant introductions. Photograph copyright 2004, President and Fellows of Harvard College, Arnold Arboretum Archives; all rights reserved.

PLATE 15. *Above*: *Ulmus pumila* L. planted as a shelterbelt in Nebraska. A shelterbelt is a formal typology, but since it is composed of species, it displays behavior, domains, and contextual variability as plant life adapts to the given soil type, artificial spacing, and available resources. Photograph courtesy of USDA archives. *Below*: Dryland planting experiments highlighted American afforestation-focused research on growing woody roots in the sandy grasslands, requiring documentation and trials to monitor root growth. Here, the vigor of roots provides an indication for shelterbelt design. Photograph courtesy of USDA archives.

PLATE 16. Afforestation across the Taklamakan Desert, the second-largest moving-sand desert on the planet. The design typology requires artificial land flattening or compaction in advance of installing the "straw grid" configuration and drip-line irrigation. The next step is tree planting in units, a tally required to ratify global mandates. In all cases, the trace of afforestation typically follows state-sponsored infrastructure. Photograph copyright 2006 by George Steinmetz.

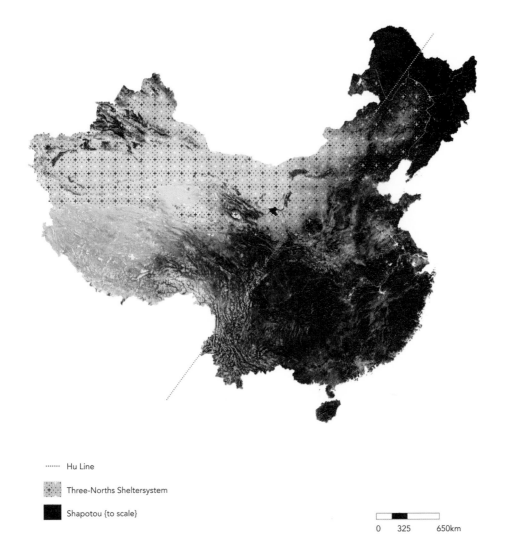

.... Hu Line

Three-Norths Sheltersystem

Shapotou {to scale}

0   325   650km

PLATE 17. Three Norths Shelter System, 1978–2050. Chinese afforestation is sustained by modifying goals in phases, enlarging control of vastly differentiated cultural and biotic landscapes. Map by the author.

PLATE 18. The Shapotou Desert Experimental Research Station lies on the fringe of Zhongwei, along the Huang He, or Yellow River. In this 2021 image, the phase 2 typology of "artificial ecosystems" is visibly connected to a rich water supply securing irrigation, unlike the areas of extensively arid regions these tests or pilots tend to inform. Photograph by Universal Images Group via Getty Images.

PLATE 19. Shapotou Station is part of a network of initiatives to support the Three Norths Shelter System. Visible are the rows of straw-grid experiments particular to the Chinese case, connected to brightly colored pump houses that convey ancient groundwater resources through miles of drip-line irrigation. Image from Google Earth, 2018.

PLATE 20. *Above*: Duan Feng Quan, a horse breeder, and his son photographed after relocation for the Three Norths Shelter System. The pseudo-ecological project of afforestation often conceals a robust social policy. Photograph copyright 2017 by Ian Teh. *Below*: A villager in Minqin, a shrinking oasis town located in the Tengger Desert. The region still faces significant challenges despite the promises of afforestation projects. Photograph copyright 2017 by Ian Teh.

PLATE 21. Irrigation supplies are a requisite to planting trees in otherwise treeless biomes. When projects are not sited close to a water source, they rely on connectivity to limited underground aquifers. Here, a crew irrigates *Populus* spp., the deciduous tree in the foreground. Photograph copyright 2017 by Ian Teh.

PLATE 22. *Above*: *Populus euphratica* is a desert poplar tree, acclimatized to the extremes of drought over millions of years. In the Tarim River basin, trees that appear to be individual are actually ageless clones, entirely interconnected belowground. Plant life adapts to limited resources by sharing supplies in the rhizosphere with thousands of other species. Photograph copyright 2006 by George Steinmetz. *Below*: The root–rhizosphere boundary. Scientists find it hard to "draw a line" between where soils begin and roots end; the same is true of where the soil ends and plant life begins. Photograph by the author, 2020.

PLATE 23. *Populus euphratica* thrives in the arid plains between various oases, river basins, and full-desert areas, tracing wet underground deposits and thriving in summer floods. Broadleaf and mixed forests are iterant and sporadic in drylands, affording support for both settled and nomadic people across centuries. Photograph copyright 2006 by George Steinmetz.

PLATE 24. The straw-grid typology reveals the *trace* of afforestation common to the Three Norths Shelter System across China, 2019. The design typology is superimposed on the landscape regardless of context. Photograph by Universal Images Group via Getty Images.

between human settlement and the climate. It is not a contour; it is a projection that univer-salizes as it divides, positioning an argument for policy that has endured for almost a century: the Chinese population is overwhelmingly located southeast of the Hu Line. The line confirms a population ratio of 94:6, proliferating hard evidence for *overcoming the line,* a catchphrase that underpins national policy.[16] Drawing a line in plan erases the accumulation of experience that underpins thousands of years of settlement, achieved in the careful amalgamation between human micro-adjustments and the macro-modifications of aeolian processes.

> Observers have suggested that every square mile of the earth's surface contains pieces of dust from every other square mile. It is true that dust is transported for great dis-tances. . . . One of the chief factors of removing dust from the air is the rain of humid regions. It is therefore apparent that much dust will not be transported into humid local-ities before it is washed out of the air.[17]

The dust that accumulates in the atmosphere is never as neutral or static as we suppose it to be.

In the same way, lines are not neutral; they include or exclude based on detachment from lived experience. Lines are drawn with similar influence along the 100th meridian in the United States, across the grasslands of the sub-Sahel, and in this case, as an oblique anthropogenic divide that does not follow a biome or a longitudinal passing. The authority of drawing a line on a map, plan, or projection also endorses the capacity to redraw the line. The population split confirmed by the Hu Line traces the experience of human settlement. Migrations and move-ments progress with a close familiarity to the shifting conditions in a thickened space between humidity and aridity. The space of this shift is transitional, not unlike a landlocked shoreline, a unique culmination between aeolian processes, plant life, and temperature. As a result, it is also the space where dust storms are abated, a reminder that rain washes dust from the air. This means that the areas southeast of the Hu Line (SEH, Southeast Half of Hu Line) are typi-cally humid, while the northwest areas (NWH, Northwest Half of Hu Line) are characteristi-cally cold and arid. If the line was thickened and shifted, it would suggest the upward-rising currents of air and the height to which dust can be carried, essentially describing the work of the wind over millions of years.[18] So the propensity toward drought in the NWH is the out-come of abrasion and deposition as it associates with below average rainfall. Such combined climatic and geological strength inhibits fixed human settlement, an association with history that is not only a narrative sequence but also the outcome of human familiarity. The Hu Line does not only suggest population patterns and urbanization potential; it traces the temporal layers of decision-making between social and biological conditions over time, the affiliation between human and plant life.

The Three Norths Shelter System (TNSS) is also called the San-bei (Three-North), or the Sanbei Shelterbelt Region, and encompasses those lands that have not been easy to settle, the territory on the wrong side of the line. This landscape is so arid that it characterizes the tension between nomadic and pastoral lifestyles. Although the TNSS lies entirely in the NWH, suggestions of the Hu Line do not appear in the literature, a reminder of how the techniques of policy select and mobilize specific forms of knowledge. In this case, arid lands cover the vast majority of the TNSS project, including Taklamakan Desert, Gurban Desert, Kumtag Desert, Qaidam Basin Desert, Badain Jaran Desert, Tengger Desert, Ulan Buh Desert, Kubuqi Desert, Mu Us Sandy Land, Otindag Sandy Land, Horqin Sandy Land, and Hulun Buir Sandy Land. There is a marked difference in soil composition between desert and sandy lands, due to thousands of years of sedimentation and accumulation, the substratum of human life.[19] The landscape defined by these twelve drylands presents an ongoing transformation of fine material that continues to move across grassy, dry, and sparsely populated ground, depositing *gobi*, or gravel deserts, while further downwind sandy material forms *shamo*, or sandy lands.[20] The drylands of China generally stratify progressively from northwest to southeast in transitions between the variously mobile particulate of *gobi, shamo,* and loess soils. The range of drylands suggests that TNSS is positioned to overcome the Hu Line by amalgamating vastly differentiated biotic landscapes.

In China, afforestation and agriculture become powerful allies in the national mandate to overcome the Hu Line, as the global concern over conservation dissolves into restoration techniques and technologies gain attention. At the start of the project, antidesertification campaigns became the primary catalyst for tree planting in grasslands. The formation of the UNCCD and the first map of desertification in 1977 only facilitated exploitation of the vast untapped frontier of the Sanbei region once the threat of drylands was catalyzed globally. The effect finds spatial evidence as suburbanization, intensified agriculture, and resource extraction reiterate across the Hu Line.

The terms of UN-mandated afforestation enable state policy to push even further beyond the line into deserts and drylands, shaping the requisite need for rail and road infrastructure. In this double operation, tree planting aligns nicely with the national plans for increased urbanization, which foster more technical prowess and refined expertise. To explore the specificity of evidence, it is useful to explain how research becomes global capital for justifying ultimately redundant operations. Notably, the mundane task of tree planting is elevated to an experimental undertaking, satisfying the criteria of research as pilot projects are quickly mobilized through standardized technique across millions of hectares. In emphasizing the speed and the spread in the case of TNSS, scale can demand something quite different from augmentation, resolving the dependence on units. As exemplars, a desert research station and a rural distribution nursery highlight the remnants and erasures of each attempt at afforestation.

VALUATION OR COLLABORATION

Valuation estimates worth. It is a noun that summarizes the degree to which something is both useful and appreciated, such that it is an expression of worth without clear boundaries. Valuation is distinct from *evaluation,* which is rooted in assessment and estimation. The misunderstandings between value as an emotional and temporal behavior and value in regard to worth and usefulness elucidate the disparity between plant and human life. Plants accumulate value in exchange with context, including human life. On the other hand, humans use value in relation to anthropogenic resources and economic conditions. Plant life values the associations that produce context, including human care in building worth, because plant associations often contain individuals belonging to different life forms. Humans value plants for their prodigious productivity or tolerance, a value that lies in isolated, individual attributes that are not at all shared. This discrepancy is exemplified by afforestation, as the techniques of valuation adhere to reductive categories that pin "success" against "failure," completely disregarding the temporality of adaptation in drylands. Depending on the cultivator's goal, activities and species are modulated contingent upon how "success" and "failure" accrue, though in both cases humans exploit particular biological characteristics. The propensity toward singular, material worth endures in each case of afforestation, as thousands of years of iterant, physical adaptation are subsumed in order to accumulate human triumph.

Afforestation is caught up in its association with value, because planting trees is always considered an environmental "good." Valuation is hinged on material evidence, the data or proof required by the collective imagination because industrious humans tend to pass on experience through formal verification. For instance, fragments of the past register the existence of previous cultures, and relics confirm the treasures of religious beliefs. Knowledge travels through the evidence accumulated by artifacts. Relationships and tradition do not fossilize. This raises questions of how we inherit the nonmaterial past. Another way to put this is that nonmaterial evidence is left out of the process of environmental valuation. What is *left out* of the fine attunements of plant life is addressed by behavior, a description of value beyond function and resource.

Human appreciation of worth also tends to augment with age metrics—champion trees and centennials—making youth less significant and therefore more disposable: the bole radius of a giant sequoia, the extraction of genetic heritage in monumental olive trees, and the physical fences that attempt to enclose millennial oaks. In the study of plant life, age is referred to as senescence, a process rather than a linear trend. Standing in a forest, billions of microorganisms and rootlets are invisible to humans, despite the fact that the extents of collaboration far surpass vulnerable crown spread. Following suit, models of conservation transform the past from being a fertile ground in which to cultivate the future into a collection of items to be admired, seemingly crystalized in time, much like an artifact locked in a museum or an animal held in captivity.

Plant life survives through collaboration. It mingles slowly with others, amassing continuity and association. Collaboration is defined here as a coalition of forces. For instance, plant life is a vital link between humans and their land. Rather than a bounded spatial setting, plant life helps define landscape by the *time* it takes to ascribe significance to the next generation. Through this lens, landscape is more than ground, and more than culture because it accumulates relationships: between plant life, microorganisms in the soil, the climate of the atmosphere, and various human interactions. Relationships are nonmaterial. Afforestation, by contrast, requires material artifacts and evidence to advance. An emphasis on material discrepancies highlights how valuation—absent relation—is promoted by organized tree-planting campaigns. In turn, these procedures neutralize the extant landscape.

The inclination to ignore difference is a technique of valuation that excludes more than it includes. In particular, it excludes the collaboration provided by stewardship and care in a community invested in making sure that trees are not just planted but that they grow. As the Majjia project in Niger elucidated, planting units is not the same as growing trees. Growth is a temporal attribute of valuation that resists the material evidence so critical to assure funding mechanisms. Each lasting project remains linear and invariant to the nuances of time that engender landscape transformation, discounting the less-recognizable associations between human and plant life.

When a tree is planted, it is not only inserted into the living ground; it is introduced into a social context: the context of relationships and traditions. In much the same way that the "success" or "failure" of a project is reductive to plant life, human relationships are not easily refined though an agenda that prioritizes the "local" over the "global." The act of planting a tree, in contrast to how trees sprout spontaneously or propagate opportunistically with the help of wind, water, humans, and other creatures, establishes a very different commitment between species. Plant life behaves with a kind of regard and appreciation to human care and stewardship, a connection that aspires to a mutually beneficial, shared outcome. When plant life is reduced to bureaucratic planting statistic, it shifts from being a living link between species to being a fixed, standing resource. Because they cannot establish a connection with humans and other creatures, plants will seek other relationships altogether, pursuing new associations in the soil, in the atmosphere, or across the ground.

Organized tree-planting projects, supported by isolated nursery criteria, encourage farmers to remove any spontaneous plants that sprout up in calculated forest plantations and shelterbelts. Planters are encouraged to cull the extant landscape. The existing soil profile is also secondary to the endeavor. For instance, inoculation is commonly added to potted trees in order to facilitate the uptake of water in impoverished soils.[21] Inoculation is an additive process, whereby bacteria and microorganisms are injected into soil before planting. Ultimately, when the tree is replanted as a tool of the project, the inoculated soil is also inserted into drylands. Here, cultivated plants that proceed with human nurture—technology, irrigation, inoculation—are distinguished, while others wither trying to find new relations when the tree planter moves on. Each technological step leaves traces beyond the tree unit.

A consideration of time includes the extant landscape and suggests the unifying thread of dryland collaboration: the effort must be jointly valued. If planting is treated as a collaboration, then it accrues multiple relationships over time and still provides benefit to each species. Projects that are asserted across vast bioclimatic differences in effect implement the vision of governmental bodies, nongovernmental organizations, international financial institutions, and bilateral donors rather than individuals in stewardship. Each peripheral agenda denies plant behavior and disallows practical knowledge, including testimonials convergent with indigenous opinion. How do reified terms empty the landscape of value?

The procedures of dryland planting proliferate through contextual indifference, when the environment is framed as simple or uncomplicated. The tactic of ignoring differences to elevate trees over other plants such as grasses divides foresters and botanists in their search for value, for instance in the criticism that surrounds the Three Norths Shelter System.

> The State Forestry Administration has been enthusiastic about tree planting and regeneration, but less enthusiastic about restoring grasslands and other non-forest vegetation. In addition, its approach to afforestation has frequently ignored differences in topography, climate, and hydrology—all of which affect tree survival—and blind implementation of the policy under these conditions may increase environmental degradation rather than improving the environment.[22]

Can the character and tradition of drylands survive the "success" of afforestation? Or, to paraphrase an ongoing debate within conservation, how can we adapt to the environments of the present rather than try to bring back the past? It is not a simple task to account for the "success" or "failure" of any planting project across vast differences in topography, climate, and hydrology. Whether referenced as a biome, an ecosystem, or a community, dichotomies do not translate easily through the temporal reality of living environments. It follows that valuation

tends to be caught up in declensionist narratives instead, especially in regard to soil conservation, desertification, and air pollution. The radical difference between an accomplishment and a
crisis is wedged between the complications of assessment and policy—or which environmental
issue is being addressed. In these mandates, the landscape is devoid of value, meaning, and tradition, enabling *blind implementation*. Any distinction between past variations and extant conditions remains unaccounted for, a negligence toward time that cannot be replaced by "phases."

Valuation between "success" and "failure" can be raised in very different ways through collaborative terms, by rejecting the dichotomy and inserting a consideration that embraces the
relationships engendered by plant life. A new intimacy emerges that includes the agency of
elements, the biology of plant life, and the forces of the environment as unexpected associates,
enabling human knowledge to accumulate more freely and more creatively.

How can conservation protect the collaborative nature of plant life? Collaboration is one
of the most intriguing behavioral characteristics of plant life. Plant biologists call this attribute codependency. Codependency is often cited in relation to archaeological evidence of
domesticated crops, or the bond between humans and psychoactive plants, for instance.[23] I
am suggesting that plants might be valued for how they accumulate relationships. A plant is
a microbiome, swarming as roots graft and share resources, attracting microorganisms into
their complex soil-building maneuvers. As a result, collaboration does not suggest hierarchy.
It suggests that our role in planetary evolution necessitates that we work with unpredictable,
volatile interference, and that humans are often very much in sync with plant life. Under consideration here is how fundamentally different plant behavior is from human behavior, which
suggests that such nuanced, messy affiliations have been omitted from our studies of nature,
and thus the intellectual future of conservation. To work jointly with others is to oppose the
trajectory of disciplining nature, enlightening our role as a planting species.

REALMS OF CONSERVATION

In drylands, every attempt at conservation exists in two realms, protection and exploitation.
Protection is that realm expressed in defense of morals, and relates to environmental security.
The exploitative is a complex of techniques and mechanisms that control resources in order to
advance. A forest can be felled, so long as it is replanted. Trees are planted, as long as they can
be counted. Famine can be alleviated by irrigation-laden croplands, so long as drylands can be
manipulated. Afforestation is the perfect development tool because it allows both realms to
coexist through conservation—protection lost in exploitation.

Agroforestry and afforestation are the most common models of dryland conservation. Plants
dimension these projects by consolidating the mandate to restore soil or otherwise protect

resources. As we have seen, it is often unclear what is being conserved since the causes are anthropogenic. For instance, rather than address the complexity of nomadism and iterant cultures, desertification mandates planting trees to promote sedentary farming. Rather than prevent the brutal plow-up of the prairies, soil conservation furthered windbreaks to protect agriculture. In the Three Norths region, the rationale is a composite of economic incentives that amalgamate urban development and resource scarcity while sustaining global mandates that weaken rural populations. In order to shed some light on the two realms of protection and exploitation, I will draw from animal studies where a similar tension emerges between *reintroduction* and *translocation*. Reintroduction targets survival when moving captive species "back" into the wild, while translocation involves moving a species from one context to another. For instance, reintroduction moves the rhinoceros from the safety of the ranch in Texas and transfers it to India. Translocation protects the rhinoceros from development in India by safeguarding the creature in Texas. The exploitation that ensues is justified under the stacks of development permits that pose as conservation measures. Exploit here to protect there.

Yet the fine print reveals that animals require a complex array of relationships to survive, not only the positive, nurturing ones. For instance, the first demonstration of the consequences of disrupting relations between territorial mammals was observed among kangaroo rats.[24] Kangaroo rat families were translocated with their aggressive neighbors, such that all "local" relationships were kept intact during the move. But former enemy neighbors were recognized in the new context, and the kangaroo rats were more concerned about the unknown threats than the known threats of the recognized neighbors. Here, aggressive relations were demoted in order to prioritize raw survival, explicating that relationships are much more complex than we can measure. It is no different for plant life. The complexities of reintroduction highlight how critical relationships are to wildlife conservation, yet "successes" obscure how difficult they are to maintain. Although outcomes that fail to value relationships have lower survival rates, they persist because moving species is economically motivated by development.[25] It is in the best interest of developers to support and invest in reintroduction programs supervised by conservation agencies rather than refocus development. Rapid growth in the field of reintroduction biology includes intricate monitoring systems in order to establish standards, offering an alternative to complete habitat destruction. In this way, specialization grows and development spreads, a predictable outcome of a human-dominated epoch.

What thousands of animal studies have shown is that the conservation of biomes is more critical than conservation of individual creatures.[26] Each progressive understanding in anthropology and zoology strengthens conservation by reinforcing the value of relationships that are already in place, in order to avoid captivity altogether. Debates are raised concerning the success of conservation projects only when the ability to re-create ecosystems is demonstrated, as specialists move to measure the success and value of an ecosystem, destroying the motivation to protect them. The novel challenge of twenty-first-century conservation is found in valuing indigenous ranges rather than economic development. The parallel with afforestation is that forests cannot be designed, whatever the measures. Individual plants do not need protection, as much as the relationships they rely on within their habitat. Unit counts cannot replace relationships, and global deforestation in humid lands cannot be offset by tree planting in drylands.

The equivalent of biological translocation is called "assisted migration" in botanical terms. Assisted migration is a strategy led by foresters and other land managers as a response to the climate changing faster than the capacity of plants to migrate or adapt.[27] This includes moving old-growth trees but also involves specific testing with replacement species by matching seed lots, the first step in measuring the accuracy of seed germination, with outplanting sites the second step in securing predictable units. Tools include industrial monitoring to assist nursery stock transfer, seed databases to map potential habitats, and software that identifies genetic degradation. In this matrix of branded species, forest trees are most often emphasized because of their economic value. A prominent example is the decline of productivity in lodgepole pine (*Pinus contorta*), as replacement species ponderosa pine (*Pinus ponderosa*) and Douglas fir (*Pseudotsuga menziesii*) are integrated into timber operations.[28] The attempt to mimic the conditions of an ecosystem excludes relationships in the forest, especially those found in the rhizosphere. This is an example of neglecting the influence of context on the character of collaboration between organisms. Consider that when a female pinecone drops from the upper canopy and finds its way to the forest floor, it partially buries itself and awaits release of its seed. As we have seen, seeds do not all compete with one another to put down roots; this is observable to anyone paying attention to these first fifteen centimeters of the soil or to seed behavior. Unlike forests, the patterns of dormancy in grasslands and true deserts represent net loss, and low abundance to forestry agencies and development organizations trying to overcome treelessness. While some practices aim to move away from "saving" the products of nature to embrace the mobile processes of the environment, plant life remains in an impoverished middle ground, as explicated by the acts of afforestation that continue to plague drylands. If dryland biomes cannot be appreciated for their treelessness, perhaps they can be appreciated for the future forests that lie under our feet.

Thus, the great loss of diversity being played out on the planet is charted by conservation formulas that conflate protection and exploitation. Afforestation campaigns destroy to advance, in order to secure singular profitable resources and stable outcomes that keep the present protected. I do not mean to imply that we should not conserve or protect but that our practices can be reinvigorated by paying attention to the process of temporal formation inherent to plant life. Consider that it has taken much of a century to refine the temporality of ecological succession. Succession is the term used to describe the process of change, and the sequence of variation ranging from decades to millions of years, overlapping in time and contingent on the disturbance. Succession occurs as a result of a change in conditions, a dynamic that expands to include humans as an agent of disturbance rather than a witness. Ecological change is spread across vast timescales that emerge between changes in the climate and the soil, because human life evolves in the boundary layer between the earth and the atmosphere. In turn, plant life adapts at different temporal and spatial scales that do not always align nicely with the immediacy of planting a developed woody root system into the fibrous, arid ground, an artifice of ecological stability destined to create more problems than it solves.

As we have seen, woody and herbaceous plants typically develop vastly different associations underground. Woody plants tend to fuse with extensive underground mycelial networks as they physically penetrate root cells; this is the root–rhizosphere boundary described earlier. In contrast, herbaceous roots are not changed by their relationships, as fungi form symbiotically with one another in the rhizosphere instead. This is the difference between ectomycorrhizal and endomycorrhizal mycorrhizal fungi, the alteration between external and internal cooperation within the rhizosphere.[29] In heavily manufactured, imported, or otherwise deprived soils, mycorrhizal networks are usually absent, requiring inoculation. When inoculation is paired with tree planting, preference is given to the woody roots, thus omitting any improvement to the conditions that might support herbaceous spread. Techniques of afforestation omit all signs of life outside the singular ambition, failing to acknowledge the temporality of the existing soil, the extant plants, and the relationships between the land and those living off of it.

This chapter introduces the TNSS through conservation to express that despite a century of progress in conservation, very little has changed across drylands. Afforestation persists through conservation rhetoric, as accomplishments are registered by amassing units and acreage, as statistics proliferate the margins of "success." The frontiers of planting in the U.S. shelterbelt claimed 4,099,000 acres under forest and range management using numerical persuasion with headlines such as "2,000,000 seeds collected."[30] The trend echoes in claims made by the SFA, as China gains international attention for its goal to plant sixty-six billion trees by 2050.[31] Substantiation accrues even before the project is planted; for instance, the FAO's Building Africa's Great Green Wall catalogs 780 million hectares of coverage, to affect 232 million people, because 21 percent of the core GGW area is in need of restoration.[32] Yet each parcel, unit, and acre represents the pretense of planting a species that either does or does not adapt to the project. This raises the question of what afforestation is funded to restore, conserve, or protect. If the insertion of tree units into treeless drylands breaks up the long-term biological association between organisms as a means to create new ones, then valuation must concern itself with the novel associations that are created in and between the human and plant communities—the relationship, not the unit. The omission of extant relationships sharpens the suggestion that afforestation schemes are incentivized as a means to secure more federal, state, or governmental control, not as a means to counter poverty or environmental degradation. How might conservation include plant life in order to gain more influence on human affairs? Can plants ever be enthusiastically—and not cosmetically or industriously—brought into practice?

8

Three Norths Shelter System

THE THREE NORTHS SHELTER SYSTEM (TNSS) is also referred to as the Sanbei Shelterbelt Region. It is one of six major developments overseen by the Chinese Forestry Administration (SFA): the Program for Shelterbelt Development along the Middle and Upper Yangtze River, the Coastal Shelterbelt Development Program, the Farmland Shelterbelt Network of the Plains, the Natural Forest Conservation Program, the National Program against Desertification, and the Taihang Mountain Afforestation Program. Chinese policy also supports the implementation of two of the world's largest dryland projects: the Natural Forest Protection Program and the Sloping Land Conversion Program. Each development plan is determined by a range of conservation policies (also termed ecological restoration or rehabilitation programs) to address an equally impressive number of environmental issues: soil erosion, land degradation, flooding, desertification, salinization, groundwater scarcity, biodiversity, and most recently anthropogenic climate change. The accumulation of issues is neatly bundled by policy, turning a project into a region. The TNSS is the largest unified design project the world has ever seen, and the most extensive act of horticulture embossed on a terrestrial biome—the landscape of afforestation.

This chapter explores how afforestation endures through design, even as global crises shift over time. In the case of the TNSS, the outcomes are expanded by fine-tuning a phased approach, liberating the project from its local and social dimensions. Although the unrestrained scope complicates analysis, isolated figures elucidate its scale: Leading up to 2005, 2.2 billion hectares of plantation-based afforestation were established. An additional 30 million hectares were seeded by aerial means. But the overall survival rate in 2006 was estimated at only 53 million hectares. Current reviews amalgamate total expenditure at more than $23 billion, although the initial project estimate was only $300 million.[1] The disparity of millions of hectares and millions of dollars is not easily explained. Neither is the loss of so many millions of "established" plantations. There is just so much contradictory data and so many thousands of reports spanning forty years that the project resists complete accounts. In the knot of phases, the TNSS emerges as the only comprehensible manifestation of the exaggerated effort to "scope" the entire Sanbei region.

Some quick clarification between the project and the scope is in order. Scope is a code word that refers to both the target and the purpose of a project, conveniently serving expansion and enlargement. Scope hides the actual scale of the project. In design language, scope is a kind of currency because it masks economic outcomes with ecological ambitions. It is the voice of the consumer: subjective, autocratic, erratic. Scope is a clever way to grow a project. As a result, scope maintains a commodified environment, blurring the boundary between land-based knowledge and specialist ambition.

> The local people have, since 1978, set the alignment, width, depth, density and varieties of both the primary and secondary forest belts, built shelter belts in narrow or netted shapes, which are well linked up with road, riverside, canal, and village shelter forests, economic forests and scattered trees in farming areas as well as windbreak, sand fixation forests, fuel forests, soils and water conservation forests and watershed shelter forests in the periphery of farmland. Such multi-tiered farmland networks have brought about notable ecological effect and economic returns.[2]

The difference here is that the scope and the project are not quite the same. This scope transforms active decision-making into static, replicable concepts that are legitimized by participation or community involvement. Nothing is lost, but nothing is interrogated. Because the creeping scope feeds the TNSS project, it subsumes regional scales quite unlike any other project of afforestation. The suspicion arises that the reality of the everyday is eradicated in order to give power to the cravings of the economy. What remains is found in the description of *trace,* the elaboration of quick-fix afforestation.

The elasticity of scope transforms every ecological "problem" into an economic opportunity. But scope cannot operate alone; it relies on the persuasion of phases. Allocating distinct phases enables the SFA to synchronize the project, shifting technique and changing course to keep up with demand. Each phase illuminates how the project adapts to crisis, alternating between ecological (public) and economic (state) concern, an important pairing in the rise of afforestation campaigns. As an ecological proposal, reforestation, afforestation, and eventually agroforestry swap roles as long as trees are being planted. As an economic proposal, the project absorbs nationwide objectives to invest in technology, biology, and forestry. All specialists are united by their efforts to improve dryland planting, in solving the problem of fallow fields, irrigation failures, and growing sinkholes. The evidence of increased specialization accumulates as product and process begin to merge through biotechnology, tampering with species, designing new cultivars, and retooling species lists. The force of phases washes away the context of humans and other beings.

Phases also help break down the scales that scope devours. For instance, at its outset in 1978, the TNSS was divided into three phases. The first phase was inspired by the UN Conference on Desertification, under the rubric of "dryland development." Subsequently, the project extends across four phases, with a probable fifth forthcoming.[3] It is not hard to imagine a sixth or a seventh as funding shifts the objective by profiling the need for increased "forest cover" by 2050. Cover is measured in millions of hectares, because billions of number units no longer resonate when accounting for 42 percent of the continent—a scale comparable to Austria, Belgium, Czech Republic, Denmark, Finland, France, Germany, Ireland, Italy, Luxembourg, Netherlands, Norway, Poland, Portugal, Spain, Sweden, Switzerland, and the United Kingdom combined. With such a comparison in mind, is there anything more radical than hiding a continental afforestation project in phases? The next crisis in line is evidently climate change, as scope expands to catch up with the times.

Individual phases typically nest another set of phases, adding to the difficulty of reading the project as a whole. The nested phase approach is a coercion that scripts analysis; in other words, each phase can more easily be evaluated as a means to improve the next phase, which necessarily raises concerns about what is being assessed. As we have seen, spatial (linear) scale is controlled by the economics of typology. Examining typology helps articulate how technical construction details often ignore the extant landscape. Taken together, the phased approach supports a long-term investment in typology, endlessly upgrading each construction detail to suit economic ends. Afforestation advances as an ecological necessity backed by economic assessment. Moreover, the endeavor is validated by claims of local support and engagement. In the second phase of the TNSS, phases compile the complex of ecological issues into a cluster, an approach that integrates systemic issues in order to appear singular. Keep in mind that singular issues tend to be addressed with singular solutions.

Ecological issues are not singular. They cascade, accumulate, and extend as they recede. As we have seen, this is one of the motivations for *index,* as specialists attempt to reason with the predictable virtues of our planet only to disregard others. In China's Sanbei region, ecological issues swarm. In the 1990s, a record twenty-one sandstorms and five droughts pulsed across the landscape, posing accumulative pressure on towns and villages.[4] Sandstorms accumulate drought, loose sandy sediment, and extreme winds in a period of relatively low precipitation. Of course, sandstorms are blamed on desertification and poor farming techniques, outlining seemingly singular problems in order to prescribe afforestation. As concern over resource scarcity (ecologic) and the promises of secure income (economic) emerge, villagers are mobilized by nationwide support for the TNSS. Because it is a tree-planting project, it empowers afforestation, and the campaign expands across multiple dryland provinces. In this second phase, families and villagers are relocated and forced to abandon pastures to make way for the project. The urgency is not actually the vulnerable dryland condition or the villagers but the airborne particulate making its way toward the coast, and over Beijing.[5] Personal accounts accumulate of families forced to leave villages across the Kubuqi Desert: "If you didn't move they would demolish your house."[6] The social impressions of afforestation are entirely obscured by the urgency of policy mandates to plant trees.

Because the TNSS is a tree-planting project, it conveniently integrates ecological concern backed by economic validation. The local population, forced to relocate, now labors to plant trees and keep infrastructure sand free, while trees reassure global anxiety and the pressure to meet UNCCD mandates.[7] Because it is a tree-planting project, a new phase is born to convey funding mechanisms, stepping up groundwater draw and adding irrigation, automation, and biotechnology. Phase three begins, and scope extends to garner more "local" support, improve design typology, and augment species lists. Because it is a tree-planting project, no one is asking any questions. Afforestation is superimposed on the landscape, sanctioned by national policy. Global "environmental" awards extend the reach of power, the reach of each tree unit.[8] This is not fiction; it is shorthand for the accumulation of fact nested in state reports, academic papers, and firsthand accounts. The shifting frameworks confirm that global conservationism is in the habit of depleting existing relationships and replacing them with new ones. This superimposition rehearses violence to plant life that relates land-based knowledge to the specificity of agitated biomes. Projects blanket dunes, expanding without asking what species is being planted, where, and by whom.

EXPERIMENTAL STATIONS

China's drylands cover 37 percent of the country. The conditions of the territory are multifarious but characterized by a monsoon climate: a harsh temperature gradient that yields very long cold winters; short, hot summers; and very little in between. The stunning difference in temperature creates soils accustomed to the regularity of strong wind, soils that are mobile, loose, and erode effortlessly. Such mobile soils are hard to pin down, and they invite similarly iterant species into their consortium. Dryland soils are often called "highly erosive," which is really only a problem in the face of human settlement. Accordingly, drylands are not particularly hospitable for humankind, and even less hospitable for fixed-culture development. In the relentless scope of the TNSS, the tendency to pacify soil, which might be called a kind of impersonation, a kind of camouflaging of authority, and hence responsibility, is most visible in the extensive straw grids, installed to appease slope and slow surface particulate.

The straw grid goes hand in hand with afforestation's most persistent themes, themes that appear through the lens of *trace*: establishing experiments to confirm typology, a replicable and generic substructure.

Straw-grid experiments and installations are particular to the Chinese case. Other tests common to the project include the arrangements of brightly colored pump houses that steady the plantings by plumbing ancient reservoirs to the surface through drip line. The faith in *Populus* species, a predictable and well-regulated plant that leads each effort, is another conspicuous aspect of the project. Each experiment in typology, infrastructure, and planting specification is linked to a network of dryland experiment stations that spread across the seven provinces of the region. Experimental dryland stations act as the hub to the network of afforestation initiatives, confirming typology, designing infrastructure, and tinkering with plants to advance dryland development. A station located in Shapotou was set up solely to attend to the success of construction in hyperarid environments, meaning it was poised to mobilize the planting protocols for the TNSS in much the same way that the Bessey Nursery provided precedent for American afforestation. Although the procedure is the same, the dissimilarity lies in scope because Shapotou is only one of thousands of experimental stations.

Shapotou is a district in the Ningxia Hui Autonomous Region, located along the periphery of the Tengger Desert in the north-central portion of the country. The Shapotou Desert Experimental Research Station (1956) lies on the fringe of Zhongwei, where the Yellow River (Huang He) bends northward. The Yellow River regulates local moisture regimes across time and in season, engendering a diversity of life in drylands. The station was established around the same time as the Baotou-Lanzhou Railway (1954–58), also called the Baolan Line, a major infrastructural investment to connect the remote deserts of Inner Mongolia with the humid, urban coast. The one-thousand-kilometer track breaches the Hu Line and provides the only link from Zhongwei to Beijing. Shapotou Station was inaugurated to run erosion control trials and determine the long-range influence of airborne particulate.[9] In other words, Shapotou was initially constructed and funded to determine how to keep hundreds of kilometers of newly constructed rail infrastructure sand free. By 1978 it was inaugurated as one of the principal centers of the TNSS.

Unlike the Sandhills, Shapotou is not located in a bioclimatic anomaly. But what it does have in common is that the station was incorporated to ground the project in *experience*. Other research stations and pilot projects are reliant on Shapotou, including those in Bayan Nur, Yinchuan, Deng Kou, and Minqin County. But the proximity of Shapotou Station to both the sandy desert and the Yellow River clarifies its designation, since one of the key outcomes of successful pilot projects is their proximity to a reliable water source. In China's drylands, annual precipitation is low, but in the Tengger Desert it is less than 180 millimeters. Compare that to Central Niger, where the Majjia Project receives 500 millimeters per year. In arid biomes, irrigation operates as the only counterpoint to humidity, so proximity to a river is essential to experimentation. Experiments in typology blanket the edges of the river, explicating how the landscape is rendered increasingly "productive" through a portfolio of technical formulas. A technical formula is a replicable operation, for instance, measuring groundwater withdrawal or application, and evaporation rates can be measured on each plantation in the same way. Baseline data is critical to a measured experiment. Yet the command of replicability forces procedural gaps, gaps that remain at the station and create friction in the field because measured

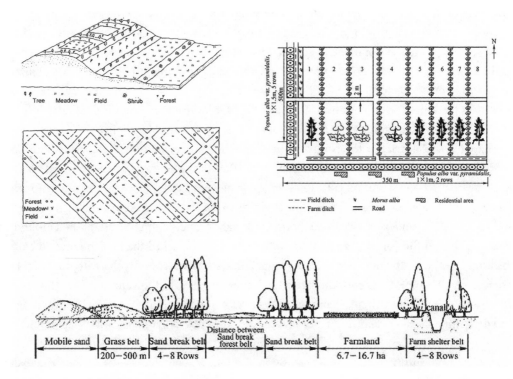

FIGURE 7. Trace reveals that afforestation survives or dies based on plant life, regardless of typology. The generalizations of typology shift through different phases as simplified drawings appear to reflect research. Images reproduced from Xiao Duning, Li Xiuzhen, Chang Yu, and Xin Kun, "Rural Landscape Ecological Construction in China: Theory and Application," in *Landscape Ecology Applied in Land Evaluation, Development and Conservation,* ed. Dick van der Zee and Isaak S. Zonneveld (Enschede, Netherlands: International Institute for Aerospace Survey and Earth Sciences, 2001), 221–31 (*above*); and Longjun Ci and Yuhua Liu, "Biological and Technical Approaches to Control Windy Desertification," in *Desertification and Its Control in China,* ed. Longjun Ci and Xiaohui Yang (Beijing: Higher Education Press, 2010), 351–426 (*below*).

experiments must be drafted into typology to universalize installation (see Figure 7). The gap widens when measures omit irrigation and root zone dynamics, complicating the success of typology outside the station.

The experiments at Shapotou are dependent on a close calculation of buried irrigation pumped into each detail from the Yellow River, a conveniently proximal water supply. As a result, testing is particular to the inclusions and exclusions of the reports from scientists and engineers employed at the station, while the rhizosphere emerges as the critical support to the project. In drylands, planted—as opposed to spontaneous—plants will simply take longer to adjust, grow, and adapt more slowly, or expire without irrigation. As *Faidherbia* suggests, trees with greedy canopies are biologically ill-adapted to a groundwater table that is eighty meters below the surface.[10] Rather, plant life activates subsurface gradients in response to low moisture content in the air.

What is left out of typology is exactly what needs to be acknowledged. Proximity to a water source is not a detail that can be omitted by a squiggly graphic arrow that points in the general direction of water. It is not a minor detail that can be solved with concrete channels and irrigation canals when the typology is designed as a replicable scheme and the scope is 42 percent of the country. The exclusion of a stable water supply from the design of typology is a disguise that creates friction in moisture-poor sites because most afforestation spreads across drylands far from water sources. What is disguised is context. What is revealed is the lie of replicability. Therefore, typology is a calculation of risk asserted through time. The risk is low in the short term, when units are calculated by vigilant global agencies, but high in the long term if *growing* trees was the intention of the project. The project is tested with sufficient supply, but installed into lands with inadequate supply. Thus, *planting* trees is the crucial character of afforestation. The relationship between success and failure evolves by omission. Planting and counting trees are the necessary contributions of dryland experiment stations.

At Shapotou, the experimental "pilot" programs responsible for tallying units are labeled "artificial ecosystems." The first step in creating the pilot involves making the ground "operable" by leveling, synthetically erasing the variegation of the dune topography and the fine, living biological crust under the surface.[11] The procedure flattens two- to twenty-meter-tall dunes with bulldozers, preparing the land for planting. Imagine a flat surface nested into a concentration of undulating topography. Next, pumping stations are installed at the periphery of the ordered surfaces, as river waters and river sediments are conveyed to each flattened plot, pumped through thick, black, plastic tubing. As the rich, watery sediment is mixed into the top layers of sandy surface, it seeps and mixes into deeper horizons of the soil. The water saturates the flat plot, and holes are dug at regular intervals in preparation for planting. Typically, nursery-grown trees are inserted as a next step, although drip-line irrigation might also be laid depending on the experiment. Either way, the roots mingle with the fine stratification of river-ine microorganisms, pressing into the artificial substrate to find more moisture.

Drip-line irrigation is a buried network of plastic tubing perforated to dispense water directly to the substrate and the root zone of plantings. The outlines of winding drip line trace the needs of blooming suburban plantings and verdant golf courses. In dry, hot climates and in depleted urban ground, drip line provides the most effective means of irrigation because it significantly reduces loss through evapotranspiration. In other words, submerged irrigation has localized effects on plant establishment in soil that is already heavily degraded. But drip line installed along the entire profile of the Baolan Line demonstrates that a localized gain cannot be overlaid with a regional strategy. Linear scale breaks down. Pumping water from limited underground water supplies creates a composition of dependence whereby the survival of the tree unit is reliant on a series of perforated hoses, the product of erosion-control exper-imentation. But the most common setback of this "technology" is that perforations tend to get clogged by particulate matter. What are the risks when a 0.0625 millimeter particle can corrupt the system? Considering that *dust* is unavoidable in the desert, are standards being developed for the territory or as a mandate of global environmentalism? Typology ignores the material-ity of the landscape and generalizes the patches and trajectories of its formation. The cascad-ing effect is that when the water stops flowing, the tree selection fails, and the dust and sand once again reclaim the space of defunct infrastructure. The proposition only makes sense as rhetoric, a static feature of authoritative conservation that requires statistical evidence to jus-tify its scale.

The demands for a tally of units does little to account for the long-term health of the plant, its survival, and its dependence on irrigation or the confidence of regional villagers in such experimental injunctions. Armed with a well-defined question, the plant is inserted into the pilot to confirm typology and distribute numbers. The plant is a tool or an afterthought to the more important work of connecting irrigation tubes, grading mobile dunes, and stabilizing the shifting surface. Singular achievements accumulate, prioritizing short-term surface stabili-zation rather than long-term depth relations.

The pilot rehearses the ecological and economic swap of dryland development. Turning dryland soils into a series of linear estimations suggests the technical generalizations that make powerful claims over the landscape. But they can be deciphered. To extend the previous exam-ple, consider the introduction of silty sediment from the Yellow River. Trees acclimatized to aridity would prosper with the simple addition of water, but the silt increases soil fertility, a welcome extravagance for diverse annual yields. In other words, the addition of silt reveals that the pilot is not funded for afforestation it hides as a tree-planting initiative. Rather, the design suggests trees are only inserted to count units and secure fixed agriculture, the resource to eventually cultivate arable crops.[12] This is a critical distinction as scale accumulates because leveling is an indication that the afforestation effort is actually only the first step toward securing agricultural lands, cheating nomadic livelihoods with the promises of afforestation.

The landscape that ensues from pilots, tests, and trials is recognizable in the aerial outlines of dampened fields, the layers of rows and lines associated with planting, and the diverse scatter of bounded infrastructure: test plots, irrigation hoses, and pump stations. The standards of each phase are layered in the deposits of brownish green and the gradations of greenish brown, suggesting the cycles of funding and research that sanction groundwater withdrawal. Once withdrawal stops, the once-green belts fade. The remnant landscape is variably chaotic, inefficient, cleared, bounded, scarified, and eroded. Opposition is registered in the drifts of particulate that travel across the surface of obsolete typologies, while encounters with plant life can be read in the catalog of planted grids that extend in every direction, transforming the long, meteoric history of the desert into bounded fields of obsolete technology.

THE DRAW OF SHELTER GRIDS

The first phase of "artificial ecosystems" in and around Shapotou was met with mixed outcomes due to the disparity between isolated typology tested at the station and the varied character of installing typology across the landscape. In conversation with scientists from Shapotou, the distinction between irrigation technology and abundant draw was described as "poorly understood."[13] The lack of understanding is attributed to the first phase of the TNSS, a decade contoured by hundreds of thousands of rows of recessed plastic drip irrigation. In this phase, farmers, herders, and families across Shapotou were "encouraged" to settle on state-owned territory, and "set the alignment" of their shelterbelts and croplands under the assumed protection of afforestation.[14] Each mile of black drip irrigation hose connected thousands of irrigation-dependent farms. Farms were lined with trees following the "shelter-grid" typology, while surrounding dunes were regulated by mechanical and hand planting, in straw grids, clay breaks, and ditch-planting typologies. In this example, the limits of groundwater supply are invisible to the schemes of global connectedness.

Meanwhile, the geography of drip-line irrigation is obscured by its ordinary status, a mundane construction detail, rarely if ever mentioned in reference to design projects. As we have seen, the complication is simple enough; any planting typology must be plugged in to a steady water source. In drylands, gaining access to a river is either unreasonable or requires miles of supplementary ditching to create canal infrastructure, another hidden construction detail omitted by typology. These first-phase standards established by "dryland research stations" such as Shapotou led to "second-phase" protocols, as the SFA was forced to find ways to control draw and reduce the negative associations of abundant draw. While the specificity is nested in spacing and species selection, the physical outcomes of abundant draw induce land-based effects that are very real, and often exceedingly dramatic.

Sinkholes typically appear when groundwater is sucked dry, affecting small wells and the communities that depend on them. They also leave large craters in the desert, a wide void-like hollow that almost looks like a huge drain consuming sand. Scholarly reports and internet searches abound with accounts on the muted water supply across China's drylands, dissipating blame on global climate change, for instance.[15] Yet most reports keep fifty years of pump-and-draw irrigation out of the calculation. Without artificial irrigation, vast grids of even-age plantations do little to support plantation crops, and farming activates. A case not far from Shapotou Station charts the depletion of the Minqin oasis, between the Tengger Desert and the Badain Jaran Desert.[16] According to Lu Qi, a desertification specialist at the Institute of Forestry in Beijing, uncontrolled irrigation fueled the expansion of both deserts near Minqin, such that more than ten thousand people had to vacate their farms within a one-thousand-square-kilometer radius surrounding the town.[17] Resettlement in another area will only cause the issue to travel along with the relocation, while funding "water conservation techniques" along the way. Agriculture and overgrazing are typically blamed for "land degradation" without acknowledging the incentives of settlement and the water draw that follows. Ordinary drip line disrupts the image of dryland crop plantations, as research encourages the wide adoption of "suitable" agricultural practices, once again blaming the small-hold farmer.[18] Anthropogenic desertification is cited as the cause, pointing out that improper management techniques are the leading cause of ecological deterioration, which impacts economic stability. The fact that the regional water supply, limited by an underground aquifer, is extracted and exploited to irrigate the surface of the desert evades blame, which is precisely what makes it political in nature. Well-funded, short-term irrigation based on a push for continual growth leads to regional instability. The Minqin region is scarred by shelter grids, including some of the largest forest farms, the failed attempts at afforestation.

The point is that there are no simple, singular solutions that do not create more problems. The proposition that irrigated shelterbelts can enhance farmland productivity is a temporal miscalculation between scalar and temporal adaptation because the woody stems of each tree are greedy and eventually the shallow crops desiccate. This is exacerbated in villages that do not have access to flowing water sources (unlike the dryland station), an omission of standards that does not appear in drawings of typology. Desert research stations exemplify the universalization of a particular technique that is reliant on the achievements of technology, without recognizing the stippled limits of context, in this case the association between underground aquifers and regional aridity. Scale breaks down in the face of elements and calculations that refuse to "scale up" because linear scale completely omits relationships. Even the term *solution* can only be examined as a means to placate short-term demands that omit in order to advance the research question. The depletion of the limited water reserves through idealistic irrigation schemes calls attention to the disparity between practical experience and technological advances.

The pilot projects at Shapotou are characterized by two typologies: straw grids and tree windbreaks. While both provide ongoing research for testing, the straw grid is usually applied to slopes- and dunes-adjacent infrastructure, while windbreaks line agricultural fields. One design appears as a thickened linear strip embedded with a smaller internal grid pattern, while the grids of windbreaks form macro-lines that enclose and demarcate cropland. Both typologies rely on planting established through a series of linear stages. The first phase calls for establishing sand barriers, created with straw checkerboards and planted with sand-stabilizing species. This phase is called the "semi-stabilized surface," using plants that are typically low and woody such as *Artemisia, Caragana,* and *Hedysarum.*[19] The second phase outlines the "development of topsoil crust," which claims to increase soil microorganisms over time—phases in phases. Phases progress because there is no comparison made to strata of the naked dunes prior to the effects of flattening, bulldozing, and inoculating.

Finally, the third stage actually plants trees. Remarkably, *Populus* sp. (*P. pseudosimonii, P. simonii, P. deltoides, P. trichocarpa,* and *P. nigra.*) accounted for 90 percent of the plantings expended during the first phase of phase three, the phase in a phase confusion.[20] In any case, each decade of research at Shapotou charts the progress of micro-grid installation, explicating a portfolio of techniques that deserve closer examination.

> Considering that the primary wind direction is from the northwest, we flattened the dune
> field using a bulldozer in October 2013 and established three experimental areas in an
> area of 200m x 100m, with the largest dimension perpendicular to the dominant wind.[21]

Because deserts, grasslands, and dune systems are immeasurable, a line of demarcation is installed to bound or limit the extents of the typology. Straw or clay is applied in tight grids across a sandy surface, in order to increase the roughness of the ground. This is achieved by erecting a temporary bamboo or willow branch "sand barrier" that acts to reduce local wind velocity.[22] Behind the barrier, wheat or rice straw is implanted in the sand in grids. The metrics for these grid installations vary across pilot reports and are now generally established at a range of one to three meters, inserted fifteen to twenty centimeters deep such that the straw protrudes from the surface ten to fifteen centimeters.[23] Consider the practical investment in a project slated to continue until 2050. Planting fast-growing, water-consumptive trees is among the most crippling programs to life in drylands, an act of slow violence on the social and biotic structure of the land. The violence unfolds in the layers of Paleozoic groundwater that is pumped to the surface, where it is exposed to swift evapotranspiration and thirsty tree units. Sandy lands are transformed by small interactions between water availability, wind velocity, and ground surface. The presence or absence of plants in any particular season creates a roughness that influences these coordinated movements between individual grains of sand and the mechanisms of the seasons. The space just two meters above the surface of the planet governs heat exchange, conductivity, and temperature, which have a direct bearing on the transfer of windblown sand.[24] The rate of transfer is entirely dependent on the size of the particle, and these micro-processes in local wind regimes shape entire dune systems in the same way that sand swirls about as footprints are made in the desert, at the beach, and on Mars.

The first phase of the TNSS confirmed that species *Populus* was an insatiable plant. It consumes water and sunlight promptly, which is why it is valued for quick growth rates regardless of the particular cultivar or hybrid. Rapid growth yields faster turnover to processing and pulping. Yet *Populus* sustains the planting statistics throughout the second and third phases, a curious reminder despite its disappointment and low success rates in relation to water scarcity. The specifications across further decades bring afforestation into alignment with the evolution of the project in the second phase as a means to secure forest supply. Presumably, China is planting "protective forests" to control dust storms and stabilize dry ground. But a closer look at which species are specified sheds light on the actual dimensions of the project, the syntax of extractive managers that firmly belongs to the wood-fiber industry: "The established net area of protective forests is 30 million hectares, with a useable stemwood volume of 850 million cubic meters."[25] The project continues along the slippery slope of state-sponsored resource extraction, as forest belts are closely related to the statistics of timber processing. Another assessment reveals more statistics that only add up to more extractive efforts: "Between 1978 and 2008, the forested area within the TNAP region increased by 1.20×10^7 ha, but the proportion of high quality forests declined by 15.8%."[26] The elaboration of TNSS into the Three-North Afforestation Project (TNAP) attests to the shift in state incentives, which are no longer concerned with agricultural security in the third phase but with the link between dust storms and forest cover. Attention shifts to the Horqin Sandy Lands of Inner Mongolia, which stretch more than forty-two thousand square kilometers in the semiarid territory north of Beijing.[27] Horqin concerns the central government because of its proximity to the Bohai region, the economic center of China, a site of ongoing contestation with Mongolians, Uygurs, and Tibetans.

The Horqin region is usually blamed for "airborne" particulate, despite significant differences in conclusions about the dust sources and emission quantities in arid and semiarid China and Mongolia.[28] Soil dust is transported and distributed across the planet in pathways from the

Sahel to the Gobi region, becoming subject to long-range transport. Particulate size is only dangerous to human health at 2.5 micrometers in diameter, also known as PM2.5, differentiated by the terminology desert dust-PM or non-desert dust-PM. Anthropogenic pollutants mingle with desert dust, complicating models but incentivizing control measures like afforestation. It is so much easier to plant trees than to stop violent forest decimation.

XINGLONGZHAO FOREST FARM

The Xinglongzhao Forest Farm in the Naiman Banner, Inner Mongolia, is a plantation and agroforestry research station that rivals Beijing in area and spread. Here, nursery trials are real- ized at the metropolitan scale, as the tradition of afforestation is reenacted through typology that not only studies planting but simulates fixed settlements. As a case in point, the project "Afforestation, Forestry Research, Planning and Development in the Three North Region of China," also known as the "009 Project," was jointly financed from 1991 to 2002 under the direc- tion of Belgium and China, and by the FAO. The details of the project cite sixteen afforestation models developed to

> meet a common, universal goal to contribute to sustainable rural development and income
> from appropriate land-uses that integrate forestry, agriculture and animal husbandry in
> a holistic and complementary fashion, within the productive capacity of the land.[29]

Breaking down typology into "models" is a concrete illustration of the encounter between sin- gle resource exploitation (wood, pulp) and complex bioclimatic environments (drylands). The chart is organized into test regions such as PP2 and RV4 and includes a brief description and objectives, ranging from mixed fodder production to wind protection, and the requisite wood production. It also includes "environmental improvement," a goal that remains undefined in reports. But attention to the "components" section reveals that typology continues to insist on the same blocks, rows, and belts that contoured the American Prairie States Forestry Project, and are likely to find a new justification across the African continent.

The sheer size of the Xinglongzhao Forest Farm challenges a reading between research and testing versus settlement and livability. The site itself is entirely closed off to visitors yet en- closes residential farms, farmworkers, and model villages. Photographs proliferate that regis- ter the different techniques and typologies of afforestation, across almost twenty-five years of experience of building farms and walls while assessing crop yield and grazing patterns (see Figure 8). The inclusion of human settlement in the nursery expresses its scale, as a nursery- station-city. Despite the decades of criticism, *Populus* continues to proliferate as the principal

| | Site | | | |
Code	Class	Description	Objectives	Components
			AGROFORESTRY MODELS	
AF1	2	Mixed shelterbelts with alternating fodder and agriculture crops	Agricultural production, fodder production, wind protection for crops, wood production, and environmental protection	Tree shelterbelts (*Populus* and *Pinus* spp.) with shrubs Hedgerows of shrubs to separate fodder and food crops (*Salix matsudana, Amorpha, Carraghana, Hippophae rhamnoides*) Fodder crops (alfalfa, *Hedysarum, Astragalus*) Food crops (maize, beans, vegetables)
AF2	2–3	Shelterbelts for fruit trees	Fruit production, wood production, and environmental protection	Tree shelterbelts (*Populus* and *Pinus* spp.) with shrubs Fruit trees (*Pinus siberica, Prunus armeniaca,* and other fruit trees)
AF3	2–3	Shelterbelts for Seabuckthorn plantations	Fruit production, wood production, and environmental protection	Tree shelterbelts (*Populus* spp.) Fruit bushes (*Hippophae rhamnoides*)
AF4	3	Shelterbelts for rain-fed crop production	Agricultural production or fodder production, wind protection for crops, wood production, and environmental protection	Tree shelterbelts (*Populus* spp.) Fodder crops (alfalfa, *Hedysarum, Astragalus*) or food crops (maize, beans, vegetables)
			PRODUCTIVE PLANTATION MODELS	
PP1	3	Productive poplar plantations	Wood production, environmental protection, and fodder production	Block plantations (*Populus* spp.) with rows of shrubs
PP2	3	Productive poplar shelterbelts with rain-fed agriculture	Wood production, wind protection for crops, and environmental protection	Wide shelterbelts (*Populus* spp.) to protect areas used for agriculture

PP3	3	Productive pine plantations	Wood production, environmental protection, and fodder production	Block plantations (*Pinus* spp.) with rows of shrubs
PP4	3	Alternating belts of different tree species	Wood production, environmental protection, and fodder production	Mixed plantations in blocks, consisting of alternating belts of different species (*Populus* and *Pinus* spp. and native tree species)

REVEGETATION MODELS

RV1	4	Open woodland with fruit trees	Almond production, regeneration of vegetation, and environmental improvement	Small blocks of almond trees in natural grassland
RV2	4	Shrub woodland with fruit trees	Almond and fruit production, regeneration of natural vegetation, and environmental improvement	Small blocks of almond trees and Seabuckthorn in natural grassland
RV3	4	Pine woodland with shrubs	Environmental improvement, regeneration of natural vegetation, and wood production	Small block plantation (*Pinus* spp.) surrounded by shrubs
RV4	4	Poplar woodland with fruit trees and shrubs	Environmental improvement, regeneration of natural vegetation, and wood and fruit production	Small block plantation (*Populus* spp. and *Prunus armeniaca*) surrounded by shrubs
RV5	4	Willow open woodland with fruit trees and shrubs	Environmental improvement, regeneration of natural vegetation, and wood and fruit production	Small block plantation (*Salix gordejevii* and *Prunus armeniaca*) surrounded by shrubs
RV6	4	Elm woodland with shrubs	Environmental improvement, regeneration of natural vegetation, and wood production	Small block plantation (*Ulmus* spp.) surrounded by shrubs
RV7	4	Open grassland with shrubs	Environmental improvement, regeneration of natural vegetation, and fodder production	Shrubs in a grid of crossed belts
RV8	4	Open grassland on sand	Environmental improvement, regeneration of natural vegetation, and fodder production	Planting of natural fences

FIGURE 8. Summary of the sixteen afforestation models developed for Horqin Sandy Lands. Food and Agriculture Organization of the United Nations (FAO), "Tackling Desertification in the Korqin Sandy Lands through Integrated Afforestation," 2002, http://www.fao.org/docrep/006/AD115E/AD115E00.HTM.

species, supported by widening the genetic base of clones to increase productivity, and a long-term breeding and selection program for resistance to pests.[30] The program also transformed the technique of planting poplar, to "fine-tune" afforestation for large-scale plantation establishment. The struggle to assert directives and catalog, model, and ascertain is further fortified by technological mechanization, as the FAO report details.

> The development of this fast planting machine (called Medium Depth Planter or MDP) has been fine-tuned to ensure that it is technically viable to operate and cost effective. Presently, the Medium Depth Planter has proven to be a cost effective way for tree planting and it is considered a major achievement of the Project in the area where human labor itself is scarce or not available for forestry activities.[31]

The same account reports how the territory is conquered by sinking one-year-old cuttings into a pit with a depth of 1.3 meters, profiting from the loose, sandy substrate. The machine is lauded for achieving excellent survival rates. Across the margins, between mounds of mobile dunes, small settlements, and seasonal rivers, are the remnants of each pit that align as a model of planting that can be replicated. The residues of each nursery trial attest to how these agendas aim to get closer to ground yet manage to get further away from it. The false closeness is a product of environmental pejoratives that escape the report and actually yield vast physical networks of grids and lines across drylands. Under the same auspices, the project is deemed a "success" on paper, but on account of the important traits of how each network, grid, and line endures, collaborates, or dies in the spaces that are unaccounted for. There is no follow-up. Because global environmentalism sanctions afforestation, it is funded into action despite the potential of local friction. In this case, national governance trumps powerless minorities armed with evidence of the *reality*.

The technological advances that mark the final phases of the TNSS are essential to the mobilization of territory both inside and outside China, as state officials ratify means and improve relations with a host of global organizations. Declarations such as the millennium development goals (MDGs) continue to bond ecological and economic models.[32] In a notable statement by Foreign Minister Wang Yi at the launch ceremony of "The Report on China's Implementation of the Millennium Development Goals," Wang asserted that China will speed up efforts to protect the environment. To emerge from global disadvantage to global leadership is a notorious Chinese story, a tale that rarely includes the temporal misalignments between plant life and conservation efforts. The distinction in how plants exploit their environment realigns the assumption that the environment is harsh or unkind to plant life. It unsettles spatially and temporally, opening up the possibility that plant life can question and impose itself on legitimate areas of human knowledge. The long-term behavior of plant life outside economy, outside "productive" models, and outside forest farms is wrapped up in environmental politics because only a few species are closely associated with forestation, afforestation, and reforestation schemes alike. In projects that exploit plant life, growth is twisted into a product of political significance. This is precisely what happens when behavior is extracted from knowledge of the plant, of its powerful movements, life history, and coexistence—so that it can reemerge as an *artifact* in the *index* of environmentalism. From this point of view, abbreviated species lists are a powerful reminder of the weakened relationship between plant and human life.

RADICAL DIMENSIONS

China proliferates planting across drylands with a suite of afforestation typologies, and nursery programs including arid-land research stations, pilot programs, tree armies, and new technologies such as aerial seeding. Growing sites include seed sorting facilities, greenhouses, field plantations, and pilot project details. The dimensions of the TNSS are radicalized across the surface of the land, visible in the marks and scars of five decades. The range of experiments inform tree culturing, turning the attention of each typology to a new global development initiative with a novel designation, agroforestry. The summarization of typology is read through charts and tables that proliferate "models" of development, interchanging objectives and components. Codes are shorthand for afforestation, as shelterbelts are delineated by a few select species and crops.

The curious interplay of scale is instrumentalized in the third phase, by establishing replicable technique, including another seemingly mundane detail: spacing recommendations. On an aerial image, the scars of typology, evaluation, and appraisal stratify the ground. Phases swell to include "models" of development, from dryland improvement to forest resource. The proof is nested in the tally of plant units that engender indexical records. Controlled sites enclose the space of propagation, obliterating seasons and modeling irrigation demand in order to provide replicable "success" stories that graft local evidence to global frameworks. Trace materializes across the landscape of experimentation at unprecedented scales. Each upgrade furthers the indifference to context.

Nurseries propagate and test plants in order to recommend standards, including seed germination rates, plant dimensions, transplanting age, soil specifications, and irrigation procedures. In the case of the TNSS, nursery supply is industrialized to keep up with demand. Analogous to the isolation of the laboratory, nurseries neutralize context in order to establish a physical benchmark or an empirical standard. Plant stock includes seeds, bareroot seedlings, and various kinds of containerized stock, grown to materialize tree-planting policy. The nursery helps keep out context. It acts as a control, disregarding any entanglement with immediate or adjacent plants. Generating benchmarks for each species plays out in much the same way that the holotype described in the first section of the book is designated as a model to compare all future plant samples. Nurseries depend on replicable technique. The only difference here is that the operations occupy extensive national, regional, and ethnic boundaries in order to garner evidence. Subsequently, nursery supply materializes across the territory, laying claim to a strictly resource-based model of environmentalism.

The routines of a given nursery are tested for success in two ways: the quality of seedlings and their survival rates. Survival is tested when seedlings are "planted out," a trade term for moving plants from the nursery to the field. Planting out moves stock from the greenhouse to control plots that test irrigation and spacing intervals. Irrigation and spacing are calibrated to highly neutralized conditions, and create a proxy to potential water supply. Subsequently, nurseries are annexed to institutionalized dryland stations and research departments that rely on the normative curves of planting out records. The data are trusted because variables are monitored and controlled, as plant records are established in a neutral, homogeneous context. Nursery standards help build supply and determine the potential scale of each project.

Throughout the current or "final" phases of the TNSS, a wave of criticism emerged in China, asserting that the attempt to afforest 42 percent of the country is not only scientifically unsound but also practically impossible. Scholars posit that the project has *not* slowed desertification, calling efforts ineffective and, in some cases, detrimental.[33] Shixiong Cao is the most outspoken voice contributing to the criticism, calling out the program as "aggressive" afforestation. In the decades that chart the rise of the second and third phases, Cao's research shows that sandstorms increased in frequency and intensity, and degraded land increased, although forest "cover" expanded, a clear criticism that targets the extractive side of forestry.

> To support wood production, which was an economic priority, >80% of the afforestation in the Three-Norths region involved monoculture planting; often, fast-growing species with low water-use efficiency were used, such as *Populus tremula L.* These monocultures typically consumed 20–40% more soil moisture than the steppe species that the trees replaced, leading to drying out of the soils, soil degradation, and greatly increased tree mortality.[34]

Cao and his colleagues describe the unintended ecological consequences of the TNSS by highlighting one of the most pressing questions underlying all afforestation campaigns: Why are we restoring grasslands with trees? Cao published remarks subsequent to his 2008 article in order to clarify his assessment.

Because of their expertise with trees and China's desperate need for new sources of wood fiber, China's Bureau of Forestry has naturally prioritized afforestation, and has been much less interested in grassland restoration.[35]

The impact of replacing biomes comes to light in Cao's statements. While policy matters, it is his attention to plant life that provides coherence. Species selection is too often ignored by the strength of global environmentalism, as officials are inspired by technology and put off by the prosaic clarifications of what species is being planted.

The technologically driven, academically funded model of research deployed in the third phase meanders into the realm of political transformation, as the country mobilizes to keep pace with accelerating population and increased globalization. Rather than rework the species selection or maintain planted tree health, a series of technological designs incentivizes additional research opportunities, including dune modeling and tree culturing—tinkering with species and soil to develop new cultivars. The fabric of political and environmental association expands through technologies, forging a link between the central government and global development agencies.

Scientists at Shapotou concern themselves with measuring phenomena associated with windblown sand, injecting sensors, radars, and traps into the sand while projecting analytics and models in order to justify much larger "reclamation" projects. Much of the monitoring is achieved despite a lack of predictive accuracy because of the high temporal variability.[36] Arguments from scientists in sandy-land regions from China to Arizona argue that the predictions between empirical measure and mathematical models remain limited. This exposes another temporal invariant, whereby measurements and models indicate considerable variability on small timescales (seconds, minutes) and linear scales (meters, acres). The temporal projection and the spatial extents of afforestation in China are so varied and the conditions so differentiated that models move the discussion to developing specificity as new crises emerge through climate change–induced interests that help offset growing criticism.

At this point, tree planting in China is entirely military and state run, having failed to gain enough "success" from incentivizing rural farming or biotechnology corporations. State-run tree plantations and forest farms gather in the characteristic ensemble of dryland research stations, nursery infrastructure, implementation techniques, typologies, and species recommendations. Large tracts of fenced land service each research plot because the main cost of desertification is still blamed on poor cultivation techniques, uneducated farmers, and rural misuse of arable land.[37] The historiography of dryland abuse is a kind of power that shapes collective memory, and the shortsightedness of global rush resolutions can replace insight with technology.

The rise of technology mirrors the rise of dryland development by manipulating the plant along the way. Notable in respect to technology are the Han people of China, for their remarkable technical developments that radically increased agricultural productivity during the Bronze Age.[38] Techniques included fallowing and manuring patterns that supported a settled reprise rather than slash-and-burn agriculture. People tended to live in pattern with the harsh conditions, which also meant there were fewer people. At the time, central Asia was made up of enclosed basins, encompassing desert plains of gravel, sand, and clay. There is so little rainfall that the rivers do not reach the seas.[39] Over time, short-lived annuals swarm with potential because they produce a more stable resource than slower, woodier species that grow less-stable foodstuffs. Such a history of selection still contours food supply. In other words, fruits, the seed-bearing structure of flowering plants, could not be stored or traded so readily as domesticated grains.[40] Neo-European productivity mixes with modern technology to produce novel settlement patterns that are simplified into "resources" that drive export over sustenance.[41] Although this is an abridged example of early settlement in the dry, cold, Three Norths region, such a potent form of progress helps frame the intensifying character of the region in terms of tree planting and its association with settled agriculture.

As a result of the slow selection between humans and grains, the world now depends only upon three crops: domesticated rice (*Oryza sativa*), wheat (*Triticum aestivum*), and maize (*Zea mays*). Farmers who specialize in *producing* crops rather than *growing* plants now solidify the dominant role of grains, as specific regions compete to lead world markets. In 2017 the United States shipped more than twenty-eight million metric tons across the world, with Kansas leading the harvested output.[42] In order to maintain an edge, exports expand and define value in arid, irrigation-dependent lands, at the same time as widespread concern over access to food persists in the prairie states. For instance, 800,000 Kansans lack access to affordable food, and Kansas is widely known as a "food desert."[43] The image that comes to mind is one of hungry farmers forced to sell all their produce. In fact, the more accurate image shows hungry, landless farmworkers who depend on a few well-fed private conglomerates for their weekly paycheck. In China, the mix of state control, global investors, military and private enterprise, and foreign aid flourishes in times of crisis. The provisions given to plantations and model farms outcompete the sovereignty of neighboring villages. Can development slow down in order to take stock of our planted planet, and process the role of plants on the character of our planet?

9

Species *Populus*

THE GENUS NAME *POPULUS* is ranked below its family but in advance of its specific epithet, distinguishing kind through similarity and difference. It is also referred to as *Populus* spp., rather than using the more common abbreviation for species, "sp.," whereby the shorthand "spp." signifies that there are multiple species and/or subspecies within the genus. For instance, *Populus* sp. would indicate that poplar has only a single genus (type), but because there are more than two species in the genus *Populus,* it is described with spp. *Populus* L. describes a group of plants that share common traits and behaviors; "L." indicates that the particular species was first described by Linnaeus. Species resonates with multiplicity yet resists plural form. The intermediary state of being both individual and multiple is a fitting framework from which to consider the behavior of *Populus.*

Species *Populus* share a number of behavioral attributes, including rapid growth rates, ease of propagation, propensity to hybridize, hardiness of extremes, wind pollination, and a capacity to produce immediately viable seed.[1] *Populus* is represented by anywhere from twenty to seventy-five species across the world, but general agreement proposes that the genus contains "approximately" thirty species, including *Populus tremuloides, P. alba, P. sieboldii, P. nigra, P. simonii,* and *P. euphratica.*[2] In more common terms, species *Populus* is also known by a multitude of tags, including poplar, aspen, and cottonwood. For the purposes of this chapter, I will refer to *Populus* spp. when I am referencing the totality of poplar as a genus but refer to differentiated species by their Latin binomial whenever possible, because each species displays different behaviors and because there are few other naming alternatives. Either way, *Populus* emerges as a plant that challenges the term *species.*

The common definition of species is simply a population of similar organisms, including plant, animal, human, and bacteria. Species are unique evolutionary individuals. From a botanical perspective, *species* refers to a distinct group of organisms with the potential to interbreed.

While the history of attempts to decipher a single definition of species is fraught with challenges, plant behavioral studies offer insight. Consider how the following statements pair:

> No one definition [of species] has as yet satisfied all naturalists; yet every naturalist knows vaguely what he means when he speaks of a species. Generally, the term includes the unknown element of a distinct act of creation.
>
> Not just animals are conscious but every organic being, every autopoietic cell is conscious. In the simplest sense, consciousness is an awareness of the outside world.[3]

In the first statement, Charles Darwin defines species through *the act of creation,* suggesting endurance and reproduction—the potential to interbreed. The second statement, by Lynn Margulis and Dorion Sagan, illustrates a reciprocal tendency at the cellular level: plants are aware of the outside world. Taken together, every organic being (species) is making choices within its environment, although choice is hardly a term that is easily related to plant life. Can conscious activities register at the cellular level?

Populus spp. exemplifies the distinct act of creation, as it hybridizes freely and crosses readily within its kind. *Populus* spp. also clones itself by vegetative reproduction (asexual reproduction), a process whereby a plant can create offspring that arise from a single organism. Vegetative reproduction enables a colony (genet) to advance by both aboveground stems (ramets) and underground stems (rhizomes), eventually detaching as new individuals that are genetically identical (clones). The terms all refer to the same reproductive capacity, although they display different "parts" externally. How plants make more of themselves is crucial for determining the definition of species, as the elusive "unknown element" invites a consideration of decisive reproduction, revealing behavior at the cellular level. Individual clones offer their genetic details to both scientists and to the wind, locating activity, seeking relationships, and stretching across the territory by sprouting, cloning, and foraging. Poplar is elucidated as an active population and an enthusiastic partner, offering reciprocal temperaments that wear, mark, map, exhibit, and express geographic ranges.

This mixed state, of being able to reproduce through cultivation or spontaneously, is exemplified by the widespread use of *Populus* spp. as a template for afforestation and a willing tree technology. *Populus* spp. illuminates that plant life is able to acquire resources from different sources and make allocation decisions.[4] Plant life balances the various demands of differentiated contexts because acquiring resources is critical to the reproductive needs of plants. The surprising behavior found in the reproduction of *Populus* spp. suggests how it exploits environmental, biological, and anthropogenic processes to make more of itself. A closer look at the array of propagative choices explains the dominance of the species *Populus* across dryland planting campaigns.

Populus spp. integrates with afforestation in two ways: First, *Populus* spp. spreads via vegetative reproduction or the production of genetically identical clones. Clonal behavior enables *Populus* spp. to gain territory through constant regeneration and continuous development through underground stems, extending the role of the rhizosphere.[5] Second, *Populus* spp. exploits the atmosphere to spread pollen. *Populus* spp. hybridizes so freely and creates so many regionally specific hybrid individuals that their extents are often called "hybrid zones."[6] Each reproductive tendency overlaps and intersects.

Moreover, *Populus* spp. is globally present and widely regarded as an ancient tree, although a singular age or range is difficult to pin down. Its prevalence is widely recognized, yet its importance is difficult to reduce to numbers, graphs, or girth. Many plants take well to either cultivation or spontaneity, but *Populus* spp. excels at both, presenting no specific preference for how it proliferates. If one tree is cloned by science, it gives rise to millions, but if a tree clones itself, it can also give rise to millions. The implication that some plants actually work well with humans can significantly rework the terms of intensified agricultural and silvicultural progress as well as dismantling the attitude that humans have forever impeded and halted biological progress. Poplar and human species collaborate well together.

CLONAL POPULATIONS, PROMISCUOUS HYBRIDS

Populus spp. is globally distributed in the temperate regions of the Northern Hemisphere, covering a range of landscapes from Greenland to Mongolia, from Japan to North America. It is defined as the tree of peoples, communities, and populations, translating directly to "people" in Roman, Greek, and Latin lineage.[7] In Chinese pinyin, *Populus* spp. is simplified to *bái yáng*, whereby *bái* refers to white, pure, and *yáng* more directly refers to poplar. This extensive range of geographic and cultural suggestions structures a reflection on extant plants and ancient forms: plant life as it appears and disappears, driven by severe climate events dating back to the Pleistocene.[8] A striking feature of *Populus* spp. in relation to humankind is its status as one of the oldest organisms on the planet. Perhaps surprisingly, individual plants like the impressive *Sequoia* (redwoods) are not the oldest trees on the planet. Rather, the plants that rank more than five thousand years all share clonal behavior.[9] Clonal plants are even considered immortal because vegetative reproduction enables plants to persist in place and move into new habitats, where they can grow dormant underground and send up new shoots just in time to start moving again.

The ability to reproduce vegetatively—to clone—is a reflection of indeterminate growth, an attribute that differentiates plant life. Humans and other creatures display determinate growth, outlined by a predetermined genetic structure. Plants grow indeterminately, which means that above- or belowground stems elongate indefinitely, an open-ended engagement with the environment. A plant's continuous growth depends on specialized cells and tissues, while development is the sum of all the changes that progress to elaborate form. Often, the form of clonal growth is expressed by sprouting, which results in the production of secondary but not subordinate trunks.[10] This means that plants are able to shed organs, create new ones, and change shape, which explains why an acorn becomes an oak tree rather than a really big acorn. The capacity to clone and sprout is an adaptive behavior so rich that it allows plant life to be split apart without dying. Clonal plants appreciate disturbance, circumventing the difficulty of spreading seed. This is how clonal populations gain territory in water-stressed drylands, multiplying infinitely across the land and agelessly across timescales.

Populus spp. is a population. Clonal populations share space. Ramets may remain attached to genet for a time, only to become a separate entity or endure as a continued extension of the plant, increasing overall size and authority. Biological interactions of this kind can overcome physical constraints, not only through decision-making but in communication, indicating that plants are able to integrate numerous sources of information about the overall status of the population along with fine-scale information about localized sections.[11] Clonal populations make choices as they integrate signals from the individual to the ramet to the clone. They do not need to spend time or energy finding a mate or engaging with pollinators, which means that the population saves reserves in order to produce more offspring. *Populus* spp. spreads quickly based on whether or not it can benefit from the movement of humans or strategize with the environment by reproducing via seed.[12] Therefore, creating clones is an independent undertaking, an assessment of the situation that enables the plant to propagate itself without extensive human care or specialized pollinators. This fine resolution highlights the insufficiency of our language to describe the messages and interactions that course through miles of interconnected plant populations.

Clonal roots are linked by innumerable forms of cooperation, overriding the image of a single tree. The affiliations between clonal roots and soil bacteria enable a vast network of exchange that unfolds across time and space in ways that scientists are just starting to appreciate. Studies in plant foraging suggest that clonal plants specialize locally to distribute resources by perceiving their environment.[13] Associations multiply as root organs fuse and divide, collecting aboveground inputs to nourish and maximize life belowground. In other words, the clonal plant calibrates itself in sections, variously riding the soil profile to collect and forage in the atmosphere or the pedosphere. Under certain conditions, clonal plants extend their relationships so far beyond the unit that they grow outside typological standards, invading fields, drawing on water supplies, and interrupting experiments. They are so good at finding resources that they tend to overwhelm other plants. This tendency has variously been described as invasive, prolific, vigorous, and even objectionable behavior, especially in relation to plantations.[14] In turn, the effort to eradicate clonal plants is as active as the effort to protect them. Thus, the magnitude and range of clonal behavior raise questions of how individuals are counted and how tree units are calculated by afforestation campaigns. Is *Populus* spp. singular or plural, individual or multiple?

Botanist Francis Hallé defines an individual plant by three criteria: First, the individual must be living, and cannot be divided without dying. Second, an individual has a genome that is stable throughout the plant (space) and during its lifespan (time). Finally, an individual must be able to tell "self" from "non-self."[15] Through these defining traits, individual status begins to fade because clonal growth is hard to count, tally, and promote. The third criterion raises the question of whether plants make choices. Hallé explores the question by examining pollen, illustrating that when pollen grains from one individual land on another, the flower will recognize "not-self" and refuse to germinate. Wheat pollen falling on a rose does not produce a new species. Yet an "individual" will recognize "self" and germinate freely when pollen is deposited. This is a significant evolutionary process called natural hybridization. A hybrid is simply a cross between species that can conceivably lead to the formation of new species. While clones are genetically identical, hybrids are crosses between plants that will generate new genes, new species. Humans artificially hybridize plants when pollen is dusted between plants, creating a deliberate cross.

According to pollen records, poplars naturally hybridize with other poplar plants, creating confusion as species, varieties, and forms seem to spontaneously appear, necessitating ongoing categorical revision.[16] When the wind picks up, pollen is carried and distributed but only pollinates on female trees that can produce a tiny, inconspicuous fruit. Most poplars are also dioecious, which means that male and female flowers are housed on separate plants rather than host reproductive structures on the same plant. Thus, poplar reproduction is related to humans and other animals because the majority of plants are monoecious, a term used when a single plant bears both male and female flowers, which grow on separate trees and bloom as an elongated cluster called a catkin. This fruit typically splits and exposes voluminous amounts of downy, cotton-like seeds that are again picked up by the wind.

For instance, *Populus alba* or white poplar was a favored plant of European settlers across the treeless American prairies. Owing to its widespread use, white poplar has "escaped" and spread from individual planting sites to reach across the continent as it colonizes open river basins and disturbed sites.[17] Notably, white poplar was limited in its ability to pollinize through wind dispersal in North America, since imported plants were female.[18] In the absence of *P. alba* males, females sought productive pollen from local poplar trees, generating such novel hybrid species as *Populus alba* x *P. tremula*, *Populus alba* x *P. tremuloides*, and *Populus alba* x *P. grandidentata*.

Therefore, human settlement and cultivation did little to inhibit the advance of *Populus* spp.; instead, it equalized the conditions for expansion. Finding themselves in new territory, *Populus alba* females made a choice to prosper through speciation. Reproductive choice helps expand their range in distinct ways: aerially as a natural hybrid between species; and vegetatively in the production of ramets, persisting after injury, break, or fire, and as surviving sprouts persist long after the original tree has died back. *Populus* spp. advances with human support, regardless of hybrid events or related males.

Distinctly, plant life does not differentiate between dispersal agents, so long as they are active. Humans are ranked with the wind. From the perspective of the plant, each has a carrying capacity. Thus, dispersal suggests that the plant is *making decisions* as it works with other plants to exploit rapid human developments, including cleared land, altered wind patterns, and other extraneous conditions, by integrating diverse environmental pathways, drawing on memories, and accelerating in turn.[19] Such decisions are readily visible in seed dispersal, the prolific spread outlined by *Ulmus pumila,* for instance. This is how plants and humans and humans and plants work together: a plant is transported by humans without a mate, and the plant uses the wind to search for another species; or a plant is prized for its straight form and rapid growth, and is multiplied with rapidity, carried by the human project; or the plant finds no carrier, and decides to tap into its supplementary mechanism and clone itself by extending itself underground. *Populus* spp. is inclined to collaborate with humans but advances regardless of the association.

1-212.I-69/55.I-63/51.I.72/58

Populus spp. is the first woody plant to have its genome fully sequenced. It is easily sequenced because of its inherent tendency to populate itself, or clone on its own. Subsequently, the U.S. Department of Energy selected poplar as a model organism in 2006.[20] Once the draft genome assembly of *Populus trichocarpa* or black cottonwood was sequenced, the decoded poplar genome was released, distributed, and published in a paper coauthored by 109 scientists from 39 international institutions and agencies.[21] Poplar is now considered the foundational plant in the study of woody perennials, leveraged to make sense of plant biology more broadly. For instance, the institute hosts a public portal that shares forty-eight thousand poplar-derived genes available for use to apply sequencing technology to other more complex organisms.

Poplar was selected for two main reasons: First, it has a relatively compact genome sequence, which offers a simplified infrastructure of growth. Its relatively simple organization also explains why poplar is one of the fastest-growing woody plant species, commonly planted to provide fast cover on cultivated land and for quick turnaround cycles in the wood pulp industry.

Second, *Populus* spp. displays some of the widest geographic ranges in extant plants: from warm, temperate climates to subarctic zones, found in ranges that include floodplains and hyperaridity. The genome is extracted to provide humanity with benefits, including clean air and water, lumber, fiber, and fuels, as an intangible code manifests as a planted tree product.[22] Unlike vegetative cloning, this procedure is appropriately termed artificial cloning.

In 2006, when the genome assembly of *Populus trichocarpa* was fully sequenced, an entirely novel species was created: *Populus trichocarpa* version 1.1: a product augmented through the biological expression of raw data such that the plant is no longer referred to as a species but a *genotype,* a set of genes that describes this version of the plant. Remarkably, the genetic sequence created a complete proxy of the plant since the entire genome was extracted. Here, scientific achievement overcomes the plant as an *artifact* and dissolves its material order into a replicable formula. Such a proxy does not mimic the plant's formal structure; it is the code to its morphology, the strategic calibration between growth and development. Prior to the full sequencing, only portions of this complicated biological code could be extracted, which reinforced an edited version of the plant as choice attributes were isolated in order to convey important properties to humans. Prioritizing branching patterns and canopy spread confirms sequencing as a formerly curatorial exercise, one that derives, edits, and mines to advance.

Selective breeding of hybrid poplars began in China in 1958. Large-scale planting-out procedures only commenced in 1970, as new "species" or hybrids were released in the Three Norths Shelter System (TNSS).[23] The species coded 1-212.I-69/55.I-63/51.I.72/58 was particularly adaptable to both the arid and the frigid expanses of Chinese drylands, instigating the first large-scale plantation experiment to afforest the Three-Norths region.[24] Over a thirty-year period, *Populus* spp. was selectively bred with the aim of obtaining a balance between high growth rates and resistant offspring. The species coded 1-212.I-69/55.I-63/51.I.72/58 instigated an upsurge in the development of poplar varieties, as numerous crosses and hybrids were designed and implemented for the TNSS. For instance, *P. DN128, P. N3016,* and *P. N3014* were cultured by Chinese scientists for particular applications in the outer reaches of the Three-Norths region. Once cast into the field, each genotype helped advance techniques for planting saplings in intensive applications, including guidance toward spacing control, soil preparation, and irrigation.[25] Equipped with the genes to persist, *Populus* spp. is translated into a tool, as it enables a control point within the protocols of other less-manageable environmental mechanisms.

The poplar trees in the vicinity of the Tarim River are the oldest poplar trees in the world, emerging with the upheaval of the Qinghai-Tibet plateau. This kind of poplar tree has existed for more than 60 million years. . . . The Tarim Basin is the world's core area of these poplar trees which cover 352,200 ha, accounting for 90% of their total area in China and 54.29% of the global distribution. The largest natural poplar trees in the world occur in the Tarim River drainage area and large areas of undisturbed poplar forests have been preserved in this region. According to the investigations of Chinese scientists, the continuous distribution of natural poplar forest in the Tarim Basin covers millions of acres, and the volume of wood reaches over 1.5 million square kilometers.[26]

The "natural poplar trees" referenced by this United Nations Educational, Scientific and Cultural Organization (UNESCO) statement are *Populus euphratica*. The species is designated "natural" because these trees were not planted. The sites where they grow do not show evidence of human intervention. Rather, the poplar trees near the Tarim River grow spontaneously. Each plant evolved over millions of years in close collaboration with climate changes and extremes. Poplars that spread and sprout in the Tarim River basin of northwestern China are ageless clones. The power of reproductive choice emerges along prehistoric timescales as the plant multiplies by wind or water while impressively moving across the different soil profiles and landscape contours. The clonality of ancient root systems cannot be extracted from a description of the river, the land, or the nomadism of its human context. So why is value placed on volumes of wood?

P. euphratica is a desert poplar, acclimatized to the extremes of salinity, drought, and cold over millions of years. It is an ageless adapter and a model species for extending afforestation initiatives in "treeless" drylands and grasslands.[27] Over the past fifty years, Chinese breeding systems have relied on artificial crosses planted across an array of landscape conditions. Elucidating, assembling, and manipulating wild genetic resources is precisely how *Populus* spp. gains ground as a quick-fix greening scheme, a fundamental tool of biomass farming, and a biotechnology for afforestation. The species genome of *P. euphratica* is extracted by geneticists precisely because the patterns of *Populus trichocarpa* version 1.1 created a data set of information that facilitates the genetic manipulation of other poplar species.[28] Artificial cloning creates genetically identical individuals by exploiting the botanical process of reproduction through extensive horticultural propagation.

Crossing different species to enhance certain significant features, such as reducing brittleness, narrowing crowns, and limiting sprouting capacity, generates "superior" poplar clones. Therefore, the gregarious behavior of *Populus* spp. is systematically removed or extracted, as interconnected clones are recast as individual units. Artificial clones are individuated in order to advance biotechnology and populate afforestation campaigns. Tree-planting units are designed to be identical, upright, and vertically straight, while their sprouting behavior is suppressed to remove any objectionable growth. The anthropological excitement derives in part from making and remaking worlds, pacifying the aliveness of others.

Consider the fine-tuned decisions of ramets as they emerge from the underground stem. Imagine the population registering spread in relation to available resources, confirming patches of nutrients underground and atmospheric inputs aboveground. Simultaneously, root cells are dividing to avoid inanimate objects like rocks. Some lateral roots wither at one end, while leaves transpire and send sugars to storage. The population is crawling along the surface of the earth, providing cues that inform arrangement. How does this contrast with regular grid spacing arrayed with individual units of *Populus trichocarpa* version 1.1? Do the units communicate? If they do, how will decisions of proximity be communicated? As long as humans ignore the aliveness of plants, we will always succumb to the lure of statistical analysis in relation to tree planting.

In this chapter, *Populus* spp. activates the articulations between tree units and plant life, challenging the manipulations of afforestation. Yet it is unclear where *Populus trichocarpa* version 1.1 falls in relation to the phased approach and the scope of the TNSS. Only total areas attest to the ongoing management endeavors, as *Populus* spp. now covers more than 70 percent of the Three Norths area, an equivalent of more than 150 million hectares, statistics so significant that criticism of the first phase declares the failure of large-scale afforestation.[29] The criticism is located in survival rates and infestation statistics, suggesting limits to genetically identical *Populus* spp. plantations. The shifty procedures of scope arise in response, as units are replaced with area calculations, pivoting the response from a dryland development program to a project to "combat" desertification.

The actual, total numbers are irrelevant and far too abundant to address across the TNSS, but studies of the first and second phase reveal the failures that combine between water supply, engineered species, disease, and infestation. For instance, *Populus tremula* monocultures are expiring due to two wood-boring beetles (*Anoplophora glabripennis, A. nobilis*), diminishing more than 400 million hectares of cover.[30] The tactic of genetic improvement yields unexpected loss of territory, as the beetle moves happily from tree to tree without much effort. Improper spacing can devastate a planting project; it not only allows insects to spread, but decreased spacing also intensifies numbers of units, putting pressure on irrigation infrastructure.

Throughout each phase, different parameters regulate the need for artificial cloning: first drought tolerance, and then wood quality improvement; in the second phase, disease resistance. The ensuing infestations prompt more culture, as resistance breeding programs direct funding toward biotechnology. Thus, the project is translated from analysis to politics; as economic inputs shift toward fighting the longhorn beetle, attention shifts from a concern for the land to concern for the project.

With each endorsement, *Populus* spp. is cloned, named, counted, sorted, and released as units into tightly spaced grids. Part natural, part technological, *Populus* spp. helps afforestation slip between economy and ecology as units designed to be tallied and counted. There is nothing unnatural about a tree. But there is something distinctly unnatural about spacing a clonal plant.

A Species of Scale

Populus spp. thrives across the landscape of afforestation at interlocking scales. The austerity of planting in lines is blurred by the mechanisms of drip irrigation, as constant water supply nourishes nonlinear growth. Linear scale is deleted by behavior, as *Populus* spp. breaks the specification by overwhelming typology, the predictions of specialists by consuming irrigation, and the confines of project extents by cloning. The capricious behavior of *Populus* spp. thrives with drip-line infrastructure, outcompetes annual crops, and eagerly sprouts, pollinates, and continues to hybridize across drylands. *Populus* spp. naturalizes beyond the political limits of the project by thickening shelterbelt rows and encroaching on farmland. Naturalized is a term that refers to the spread of introduced plants. To naturalize is to adapt. Some naturalized plants are so good at adapting to new environments that they are considered weeds. In this case, *Populus* spp. confuses the terms of naturalization. Imagine instead that *Populus* spp. assimilates, thriving through the forced actions of transplanting, as it takes to collaboration with humans. It readily absorbs environmental difference above, below, and at the surface of the ground, even when the ground is leveled or soaked. Imported plants mingle with dormant relationships in the soil, just as winds help cross-pollinate eager plants that seek adaptive connections in their environment. Hybrid pollen swarms overlap in the breeze, cross, and backcross, embedding contextually specific plants across geographic scales. The static form of "tree" is superseded as the distinct act of creation—species *Populus*—naturalizes scale.

What does "species" mean to afforestation? The enclosure of "species" as a singular fact does little to exemplify how plant life both communicates within "species" and assimilates inputs from external "species" in order to respond to given environmental and anthropogenic circumstances. Moreover, plant life integrates "species" at localized, regional, and temporal scales. In this way, the term *species* is botanically produced because it only dimensions scientific inquiry, as frequent designation is required to keep pace with the abundance and ease of fertility. Because the physical commerce of plant resources does not consider how plants are exchanging through specialized relationships, the forms of naturalization beyond the nursery, the pilot, and the shelterbelt are rarely considered. Yet the temporal scale of each seedling, root bud, and pollen grain maneuvers into daily life by building relationships with other "species." When behavior is anticipated, the reductive quality of "species" lists and spacing requirements sheds light on afforestation as an imaginary environmentalism. Afforestation poses as environmental action, placating conservation and restoration at once. There is nothing "natural" about tree planting in drylands. The enduring excitement for tree planting ignores plant behavior, and its exclusion is physically read in the "species" lists attached to each project. This is precisely nature out of sync: a landscape regulated by a configuration that only appears socioecological and egalitarian, hidden behind the image of a tree.

In the meantime, *Populus* spp. is operating with its own individual authority, sped up by massive continental wind corridors and newly cleared grounds that accelerate seed movements along the tracks and treads of human motivation. As the soils are either desiccated or irrigated, the reproductive habits of adventive plants naturalize the rhizosphere and change the composition of the landscape. Scale is encompassed by the early statements of plant ecology that "no matter how full a soil seems occupied with underground parts, examination shows that there is usually room for more."[31] The lack of scaling tools is an opportunity to pay closer attention to behavior.

The landscape of afforestation is the landscape that these plants are trying to acclimatize to, which now includes dryland stations, forest farms, and silt-laden pilot projects. Each typology displays a high level of disturbance, and *Populus* spp. is preadapted to disturbance. Once set into context, *Populus* spp. thrives. Plant life migrates along introduced, novel patterns refusing to be caught or to take a passive role in the project, tracing each railway line as a novel vector, while using paved infrastructure as a conduit. Human and plant history mingle, independent of human trade agreements, and independent of project extents or unit counts. The only suggestion of jurisdiction is found in the link between drip-line and soak irrigation.

The story of *Populus* spp. is a good place to start when considering the privilege of irrigation. Without drip-line infrastructure, the plants of afforestation fail. They fail because they grow up reliant on mechanical pumps. They also fail because sprouting behavior is systematically removed from their life history. They fail when water supply dwindles, while preadapted dryland poplar plants thrive through pulses of dryness and wetness. Among the extant plants of drylands, and those that are overlooked and undervalued, are the ancient roots and rhizomes that lay buried under the sandy surface where humans dwell. Dormant remnants are not altogether expired; they are often interconnected to younger shoots that materialize along the surface. The younger shoots are more agile, and draw in moisture from the air, transmitting humidity through the clone. Artificial irrigation artificially materializes free water for the clonal population. Irrigation instigates behavior and short-term gains but displaces long-term adaption to dryness, replacing it with short-term memories of wetness.

Populus euphratica is an ancient plant that persists without irrigation. Its temporal behavior and reasons for persistence lie buried far beneath the riverbeds and the sandy banks of the Tarim basin. Clonal extents cannot be counted, and they resist industrialized processes of wood making and pulp gathering, the tallying statistics of afforestation. *P. euphratica* endures with aridity, an adaptation that explains why its genome was sought for artificial hybridization. The poplars of the Tarim basin inspire research because they date from sixty million years ago, yet they are not individuals, so individual dates are hard to pin down. Clonal populations of *P. euphratica* are deeply interconnected stands so attuned to the landscape that they mystify time and confound "species" selection. Turning to behavioral studies helps human history embrace terms like *foraging, consciousness,* and *decision-making* in relation to plants. While not without its challenges, such terms appear when all the ordinary divides between plants and humans are exhausted, when the labels and categories are shattered and we decide to either limit knowledge or adjust in turn. Plant behavioral studies offer us a glimpse into the unexplored language of plants, extending sociocultural and biophysical contexts, if we care to listen. Rather than framing plant behavior as a product of the human imagination, the intention here is to use behavior to force thought and resist creating another blanket category.

Consider *P. euphratica* integrating information from the environment, sending chemical signals throughout an immortal clonal population. Each aboveground arrangement filters cues from the landscape and stores it in the rhizosphere. As clones mature, they learn kinship within the population.

In plants, the learning process is largely molecular and, again, the aim is to increase information flow through a defined pathway. The variation that occurs is determined by the nature of the response, but the memory of it can last for just a few seconds to minutes, hours, and then months or longer; it simply depends on what part of the cellular apparatus the memory is deposited.[32]

Such aliveness produces different understandings of the shared world. It also helps produce a different understanding of scale: *P. euphratica* decides when to increase itself in time and space above and below the ground. It does so armed with decades of inputs, with intelligence. A complex association with other plants is negotiated through foraging, consciousness, and decision-making, and beyond the given framework of ecological competition.[33] Today, competition is typically framed between plant species, for instance when an "aggressive" weedy plant takes over from a "native" species. The results place "native" plants on specialized lists that conserve, protect, and defend the plant from so-called competition. To borrow from Weaver and Clements, early American grassland ecologists, competition is a factor located as an aggregate across the landscape, a factor of the plant in context. Moreover, this context is positioned underground, since plants do their living in the soil. And there is always room for more roots in the soil.

In the soil, competition is twisted into a tentative concept because "there is no experimental proof of mechanical competition between roots or rhizomes in the soil and no evidence that their relation is due to anything other than competition for the usual soil factors—water, air and nutrients."[34] Plant behavioral scientists call this quality of communication *foraging*, alluding to the recognition of coexistence through favorable and adverse experiences. Two foraging behaviors are most commonly described: foraging for light (shade avoidance in the canopy) and foraging for nutrients (in the root zone). Properly speaking, the struggle for existence in the plant world is not plant to plant—species to species—but progresses between plants and their habitat, the latter being changed by competition only based on the demands made upon it by other plants. Since competition is just one behavior among many, remembering how plants move in reciprocity along overlapping vectors informs a durable series of connections rather than competitions. Competition is consequently a debatable consideration, and might be reframed in terms more aligned with plant life.

Do clonal plants share rather than compete? If a clone is a population, how are nutrients and reserves shared? How does the clonal root system recognize itself from others? These are questions of belonging and not belonging, and of the differences between species that can clone another and clone themselves. At first, clones travel dynamically to forage, surrounded by their own offspring aboveground, while the micro-site of the root tips acts in communication with the macro-site of clonal spread. Most strikingly, clones extend in multiple directions at once, without engaging with non-self-roots.[35] The necessity to distinguish between "self" and "non-self" is critical to foraging in the rhizosphere. This pattern of sharing is expressed in the search for nutrients in the soil, as communication between ramets is integrated to produce a developmental response.[36] According to Richard Karban: "Experiments with clonal plants indicate that individual ramets integrate information about local conditions with information about the state of the larger collection of connected ramets."[37] Therefore, the clonal root population ensures the whole plant benefits from an increase of ramets, or a patch of nutrient, as signals sense and adapt to avoid inert objects or share resources. This is how clonal populations shape and inhabit the land at the same time, rescripting linear scale. Collaboration substitutes the antagonism of competition with a kind of kinship that leads to adaptation and growth.

Roots learn to forage. Foraging means that plants can displace and remove clumps and clusters of organic matter in their path, seeking particular nutrients nonrandomly within homogeneous environments.[38] Anthony Trewavas reorients foraging in clonal roots as living in a world of secretion and desiccation, as plants generate cells that can distinguish and remember. The argument for plant intelligence is scaled to the cell but is disclosed as a territorial process:

When environmental situations change, those plants that exhibit greater plasticity can learn to successfully adapt their behavior more quickly. In itself, that implies a potential to assess possible future circumstances. Phenotypic responses are commonly slow, like colony growth.... When faced with environmental change, the behavior will thus either be interpreted as insight or an uncanny ability to predict future events.[39]

Trewavas describes an intelligence of problem solving, a capacity for reaction throughout the plant that is achieved by signals and communication. As micro-movements accumulate simultaneously, they expand the territory of the plant, without planting. Foraging behavior is a learning process enmeshed in movement from within the plant itself.

Trewavas contends that the majority of clonal plants can take discrete avoidance actions when obstacles are perceived. This is not competition; it is intelligence and memory playing out as a sensible decision, particularly evident in stands of *Populus* spp. that seem to follow river basins or travel along moisture gradients. Newly formed rhizomes are selectively collaborating to avoid patches that would generate loss, surfacing in favorable patches. In poplar clones, microscopic foraging generates territorial extents because of a highly evolved intelligence, embedded in the chemical language of cellular exchange.

The behavior of *Populus* spp. reminds us of the aliveness of plants, suggesting communication within itself, to other plants and to humans. Young plants emerge within the relationship between biological and anthropogenic dispersal, an outcome of interplay between phenomena.[40] A very different consideration of scale appears as the air is stratified by pollen grains and the ground is gripped by the energy of rhizomatic spread, all while enterprising humans are inserting more trees and tallying units. The fallout of these dynamics can be read on the surface of the land, where the plant is no longer a biological force but an economy, as ranges are augmented by the capital initiatives of scientific forestry that promote carbon trade markets, biofuel production, and timber trades. The hybrids of afforestation participate by diversifying and naturalizing *Populus* spp. As we have seen, afforestation is scripted to increase forest cover statistics without mention of deforestation practices.

Plant life acquires and defends space through growth that is often vastly extended, as deposits of nutrients might be geographically far apart when the availability of water is scarce. Such fragmented nutrition supply is common to arid and semiarid biomes, which can often be a limiting factor to spread, yet clonal roots tend to grow healthier when presented with strong contrasts in nutrient status rather than in uniformly distributed structures.[41] Information stored in the rhizosphere reveals a concealed space that creates and engenders relationships because plants communicate chemically, in a language that overlaps through concealed patterns and aerial transport. Communication in plants is a highly evolved chemical exchange. One way to think about this is that humans evolved language and plants evolved chemicals. Chemical communication means that sprouts and ramets are messaging and interacting, revealing the power of clonal plants to gain territory. It is remarkable to imagine the potentially unlimited (and immortal) advance of clonal plants, transmitted through signals across vastly differentiated contexts. It is equally remarkable that *Populus* spp. progresses because humans are a planting species.

Epilogue

Efforts to understand dryland planting do not have to confront the controversy of afforestation. But the effort of planting necessarily confronts plant life. The controversy is that afforestation forces plants awkwardly into human worlds and aggressively into biotic ones. Afforestation destroys relationships, the precious associations that enable survival. Consider the etching on page 207. It does not matter who drew it or why; it emerges from Euro–Western exploration. There is something remarkable going on in this etching: a tree is growing in the desert. If you approached the tree on foot, you might wonder why it is thriving; its canopy appears full and lush. The tree also seems alone: it is not part of a system; the dunes on the horizon register its isolation. The tree emerges sideways in a gap dug into the substrate. The gap leads to a wider cavity. The tree is far enough belowground that it did not arise independently or sprout through the surface. The tree was planted. We are witnessing collaboration between plant and human life.

The description accompanying this image is "ancient cistern." But that is an insufficient description. I imagine that someone designed this shaded cistern with great care. First, a seed must be collected, discovered, unearthed, and encouraged to sprout: a woody plant, something with a canopy. But it could not be a greedy plant or it would absorb too much of the treasured water held by the cistern. The tree should not bear a taproot, or it might break through the thin surface walls. Prolific seed would also be undesirable, as samaras would accumulate across the shallows. After careful consideration, the right plant would come to mind, perhaps aligned with effort and depth of digging the cistern. A familiar tree is visited at just the right time, so that seeds could be collected for this very purpose.

The seed is planted and guarded. After germination, it still requires protection. Perhaps the seedling is elevated, away from predators, goats, and children. Once the seedling develops a woody stem, it is deemed strong and tall enough to transplant. The sapling is brought carefully to the cistern. A small hole is dug by hand; perhaps some dried manure and extra water are mixed into the lowermost portions of the hole. A careful hand helps remove the sapling from its vessel and sinks it into the vertical axis of the gap. The soil is gently replenished and smoothed over by a loving hand.

At this point, plant life takes over. Phototropism kicks in, orienting the shoots to the sun. Cells at the root tip and the shoot tip elongate and divide. Microorganisms in the rhizosphere awaken with the moisture and nutrients provided by the small act of human cultivation. As the rhizosphere activates, the orientation of the woody stem shifts. Auxin, a powerful growth hormone, pulses to the edge of the stem, which bends to gain light. As new shoots begin to reach for the sun, the plant transcends its position and the darkness of its context.

The animated associations in the rhizosphere invite other organisms into the endeavor, as young leaves transform sunlight into sugars for the roots. Throughout the dry season, the plant expands and enlarges from root tip to shoot tip. Sometimes people check in on it, providing additional water and removing obstacles. The tree thrives. Its position is ideal, if unnatural. There is a constant supply of moisture coming from the microclimate around the cistern, extending the wet season. With few other woody species nearby, the roots extend in a horizontal and vertical mesh. The water supply is protected by the emerging canopy; as leaves emerge, less and less particulate makes its way to the water supply. The tree helps filter sand and reduce aeolian transport. Because the water is conserved, the association between human and plant life persists. This is tree planting designed with practices and relationships. Relationships refer to the associations that bind living organisms together in the world.

Durable relationships engender a different kind of power.

The relationships that plant life engenders associate with other species through connections, alliances, and correspondences that are shaped or broken by growth, dormancy, and sacrifice. Some relationships deteriorate to augment others. Relationships are expressed through behavior and create mutually beneficial results. Relationships are nonmaterial and cannot be possessed, commodified, or outlined. They resist scaling up for profit. Thus, plant life asks how deeply it can connect, not how much of itself can be measured. Different questions yield different results. In this case, the tree is not just being planted; it is being grown.

Paying attention to how we engenders relationships offers a path into practice that resists augmentation, industrialization, and exploitation. It helps move beyond singular solutions and objectified artifacts, inviting a reading of how we work with other species. The inclusion of plant life changes the frame of reference through relationships, using collaboration as a model.

Plants adapt by receiving cues from the environment in which they find themselves, an adaptation that continues at all life stages. Trees are already quite mature, well developed. Seeds are more adaptable. If the environment changes rapidly—concretized, leveled, sprayed with chemicals, infused by sewage, burned, felled, or otherwise disturbed—fitter individual seeds will retain a memory of the changes and adjust accordingly. Consider that the decision to germinate in the soil or survive through dormancy integrates an awareness of available moisture, sunlight, and nutrients at different times, such that seeds may postpone or accelerate their growth in response. Remarkably, seed response is not limited to its immediate context; seeds also draw on embedded experience. Combine this image of inheritance at the cellular level with the billions of seeds lying dormant across the landscape, riding the gales and floods of the planet and being embedded in the rhizosphere.

The world is already fully planted.

Seeds fall into two categories: orthodox and recalcitrant seeds. Orthodox seeds tolerate drying, so they desiccate well either in the dry ground or in controlled storage depositories. Not surprisingly, orthodox seeds are mostly found in drylands because they can bank on the sandy, parched soils. More than 80 percent of known seeds are orthodox, but their longevity remains a mystery since some seeds only survive a matter of decades while others appear to be ageless. Seeds survive across the planet by establishing contingency plans for an unknown future, banking on the thickness of the soil. We are superficial creatures living in a system established in layers of iterative, enmeshed knowledge of both the past and the future. There are forests waiting under our feet.

How can the study of plant life shed light on the central challenge of a planet dominated by humans? It is my hope that the three episodes in this book help articulate our parenthetical relationship to plant life; plant life is neither wild nor dependent, neither under control nor particularly out of control. I write in the hope of reviving a more equitable association. Afforestation helps shift perspective through the medium of tree planting, to see our responsibility in a different light. It is, in turn, a century of enmity acted out on the planet, in a time of so-called heightened environmentalism. The inclusion of plant behavior highlights the unique and sometimes problematic way humans inherit their past—something called history. It also points to the importance of including other organisms in advancing history, or the consequences of leaving them out.

Notes

INTRODUCTION

1. Sharif Paget and Helen Regan, "Ethiopia Plants More than 350 Million Trees in 12 Hours," CNN, July 30, 2019.

2. Aldo Leopold, *Round River: From the Journals of Aldo Leopold*, ed. Luna B. Leopold (Oxford: Oxford University Press, 1993), 155.

3. Leopold, 148.

4. Michael Allaby, "Aridity Index," in *A Dictionary of Geology and Earth Sciences*, 5th ed., ed. Michael Allaby (Oxford: Oxford University Press, 2020), 38.

5. See, for instance, Diana K. Davis, *The Arid Lands: History, Power, Knowledge* (Cambridge, Mass.: MIT Press, 2016); and Diana K. Davis, *Resurrecting the Granary of Rome: Environmental History and French Colonial Expansion in North Africa* (Athens: Ohio University Press, 2007).

6. N. J. Middleton and T. Sternberg, "Climate Hazards in Drylands: A Review," *Earth-Science Reviews* 126 (November 2013): 48–57.

7. The American project of afforestation is variously referred to as the "Prairie States Forestry Project," the "Plains Shelterbelt Project," the "Great Plains Shelterbelt," and the "American Prairie States Project." In this book, I will consistently refer to the project as the "Prairie States Forestry Project."

8. The Chinese project refers to the amalgamation of afforestation projects in the Three-North region that is commonly translated as either "Three North Shelter Forest Program," "Three-North Shelterbelt Program," or "Three Norths Shelter System." In this book, I will consistently refer to the project as the "Three Norths Shelter System."

9. Oxford University Press, *OED Online*, s.v. "Plant," accessed July 12, 2020, http://www.oed.com. For changing definitions and terms, see Joseph Ewan and Chester A. Arnold, *A Short History of Botany in the United States* (New York: Hafner, 1969).

10. See, for instance, "What a Plant Remembers," an intriguing chapter in Daniel Chamovitz, *What a Plant Knows: A Field Guide to the Senses* (New York: Scientific American / Farrar, Straus and Giroux, 2012), 113–33.

11. Eric D. Brenner, Rainer Stahlberg, Stefano Mancuso, Jorge Vivanco, František Baluška, and Elizabeth Van Volkenburgh, "Plant Neurobiology: An Integrated View of Plant Signaling," *Trends in Plant Science* 11, no. 8 (2006): 413.

12. The philosopher Isabelle Stengers suggests that a science will be called "soft" when a nonspecialist becomes interested in the same topic or is stirred emotionally by her findings. See Isabelle Stengers, *Another Science Is Possible: A Manifesto for Slow Science,* trans. Stephen Muecke (Malden, Mass.: Polity, 2018), 63.

13. See Amedeo Alpi et al., "Plant Neurobiology: No Brain, No Gain?," *Trends in Plant Science* 12, no. 4 (2007): 135–36.

14. Eric D. Brenner, Rainer Stahlberg, Stefano Mancuso, František Baluška, and Elizabeth Van Volkenburgh, "Response to Alpi et al.: Plant Neurobiology: The Gain Is More than the Name," *Trends in Plant Science* 12, no. 7 (2007): 285–86.

15. Darwin referred to these cells as the "root brain." See Charles Darwin, *The Power of Movement in Plants* (London: John Murray, 1880).

16. Anthony Trewavas, *Plant Behaviour and Intelligence* (Oxford: Oxford University Press, 2014), ix.

17. See František Baluška, Simcha Lev-Yadun, and Stefano Mancuso, "Swarm Intelligence in Plant Roots," *Trends in Ecology & Evolution* 25, no. 12 (2010): 682–83, Dov Koller, *The Restless Plant,* ed. Elizabeth Van Volkenburgh (Cambridge, Mass.: Harvard University Press, 2011); Bastiaan O. R. Bargmann, Kenneth D. Birnbaum, and Eric D. Brenner, "An Undergraduate Study of Two Transcription Factors That Promote Lateral Root Formation," *Biochemistry and Molecular Biology Education* 42, no. 3 (2014): 237–45.

18. Richard Karban, *Plant Sensing and Communication* (Chicago: University of Chicago Press, 2016), 179.

19. Marcus A. Bingham and Suzanne W. Simard, "Seedling Genetics and Life History Outweigh Mycorrhizal Network Potential to Improve Conifer Regeneration under Drought," *Forest Ecology and Management* 287 (2013): 132–39; Julie R. Deslippe and Suzanne W. Simard, "Belowground Carbon Transfer among Betula Nana May Increase with Warming in Arctic Tundra," *New Phytologist* 192, no. 3 (2011): 689–98.

20. Suzanne W. Simard, "Mycorrhizal Networks Facilitate Tree Communication, Learning and Memory," in *Memory and Learning in Plants,* ed. František Baluška, Monica Gagliano, and Guenther Witzany (Cham, Switzerland: Springer, 2018), 192.

21. Lynn Margulis and Dorion Sagan, *Dazzle Gradually: Reflections on the Nature of Nature* (White River Junction, Vt.: Chelsea Green, 2007), 23.

ARTIFACT

1. There are numerous publications to consult on the confluence of colonialism and botanic study. See, for example, Richard Grove, *Green Imperialism: Colonial Expansion, Tropical Island Edens, and the Origins of Environmentalism, 1600–1860* (Cambridge: Cambridge University Press, 1995); Londa L. Schiebinger and Claudia Swan, *Colonial Botany: Science, Commerce, and Politics in the Early Modern World* (Philadelphia: University of Pennsylvania Press, 2005); Brett M. Bennett and Joseph Morgan Hodge, *Science and Empire: Knowledge and Networks of Science across the British Empire, 1800–1970* (Basingstoke: Palgrave Macmillan, 2011); and Marie Noëlle Bourguet, Christian Licoppe, and H. Otto Sibum, *Instruments, Travel and Science: Itineraries of Precision from the Seventeenth to the Twentieth Century* (London: Routledge, 2002).

2. For instance, the consequences of the binomial system are revealed in Donald Worster, *Nature's Economy: A History of Ecological Ideas* (Cambridge: Cambridge University Press, 1977); and Wendy Djinn, *Our Knowledge Is Not Primitive: Decolonizing Botanical Anishinaabe Teachings* (Syracuse: Syracuse University Press, 2009).

1. THE PROBLEM OF PARTS

1. Dirk Baltzly, *The Stanford Encyclopedia of Philosophy,* Summer 2018 ed., s.v. "Stoicism," https://plato
.stanford.edu. For more on Stoicism, see John Sellars, *Stoicism,* Ancient Philosophies (Chesham: Acumen,
2006).

2. With thanks to Lauren Ginsberg for translating and clarifying the ancient Greek. Accordingly,
ancient Greek recognizes *orme* through a range of nouns and verbs, including an impulse received from
one another, effort, with so much zeal, and eager desire for a thing.

3. J. Donald Hughes, "Theophrastus as Ecologist," *Environmental Review: ER* 9, no. 4 (1985): 297–306.
The first use of the term *plant ecology* is attributed to Eugenius Warming, in 1895.

4. Theophrastus, *Enquiry into Plants,* Book I, 3. All direct quotations are from Sir Arthur Hort's translation for the Loeb Classical Library series, published by William Heinemann in 1916. In the introduction,
Hort clarifies that however the principle of classification arose, "it appears in his hands to have been for
the first time systematically applied to the vegetable world" (xvii).

5. Theophrastus, *Enquiry into Plants,* Book I, 27.

6. Hughes, "Theophrastus as Ecologist," 297: "Theophrastus observed that a plant flourishes best in a
'favorable place' or proper country (*oikeios topos*), which modern ecologists might term its niche."

7. Theophrastus, *Enquiry into Plants,* Book I, 7. For a more complete discussion of the Aristotelian
school, the evolution of his work on plants, and his pupil Theophrastus, see Agnes Arber, *The Natural
Philosophy of Plant Form* (Cambridge: Cambridge University Press, 1950).

8. Arber, *The Natural Philosophy of Plant Form,* 14.

9. In the introduction to Book I, Theophrastus expressly compares plant parts and reproductive
capacity to animals in order to set them apart. I speculate that this would have been a sign of respect for
his teachers, a means to set his thesis apart, and a way to answer to the query of animal comparison so
common at the time.

10. In *De Causis Plantarum,* Theophrastus describes plant physiology—including generation, sprouting, flowering, and fruiting—and the effects of climate. Theophrastus, *De Causis Plantarum,* vol. 1, Books
1–2 (Cambridge, Mass.: Harvard University Press, 1976), 90–101.

11. Theophrastus, *De Causis Plantarum,* 101.

12. A large literature on colonial trade can be found in the study of plant exploration. The following
show particular emphasis on individual plants: Londa L. Schiebinger and Claudia Swan, *Colonial Botany:
Science, Commerce, and Politics in the Early Modern World* (Philadelphia: University of Pennsylvania Press,
2005); Richard Grove, *Green Imperialism: Colonial Expansion, Tropical Island Edens, and the Origins of
Environmentalism, 1600–1860* (Cambridge: Cambridge University Press, 1995); Londa L. Schiebinger, *Plants
and Empire* (Cambridge, Mass.: Harvard University Press, 2009).

13. Linnaeus describes the ideal dimensions of herbaria cabinets in Carl Linnaeus, *Philosophia Botanica* (1751). For an excellent description of Linnaeus's instructions, see Staffan Müller-Wille, "Linnaeus'
Herbarium Cabinet: A Piece of Furniture and Its Function," *Endeavour* 30, no. 2 (2006): 60–64.

14. Lorraine Daston, "Type Specimens and Scientific Memory," *Critical Inquiry* 31, no. 1 (Autumn 2004):
155. The theme of this issue is "Acts of Transmission," which according to the editors aims to mount
a theoretical argument about the role of media in human history. Taking on "media" more broadly, Daston expands on the principal upheavals inherent to organizing nature. See also Isabelle Stengers, *The
Invention of Modern Science,* trans. Paul Bains (Minneapolis: University of Minnesota Press, 2000). While

Stengers does not explicitly define transmission through plant life, she uses it in the sense of how power is transferred between parts, or gears.

15. Michael Allaby, ed., *A Dictionary of Plant Sciences,* 4th ed. (Oxford: Oxford University Press, 2012), 372.

16. The sheer volume of information and lists produced by Linnaeus included letters, manuscripts, objects, and more than fourteen thousand plant specimens. See Staffan Müller-Wille and Isabelle Charmantier, "Natural History and Information Overload: The Case of Linnaeus," *Studies in History and Philosophy of Biological & Biomedical Sciences* 43, no. 1 (2012): 4–15.

17. Charles Schuchert, "What Is a Type in Natural History?," *Science* 5, no. 121 (1897): 636–40; Walter T. Swingle, "Types of Species in Botanical Taxonomy," *Science* 37, no. 962 (1913): 864–65.

18. In *Objectivity,* Lorraine Daston and Peter Galison project the history of scientific objectivity by exploring the volumes of images that represent biological knowledge as it was established in the eighteenth century. The most salient argument emerges in the clarification between study and object, as expertise makes itself distinctly felt in the object under consideration. See Lorraine Daston and Peter Galison, *Objectivity* (New York: Zone Books, 2007), 55–105.

19. Linnaeus included species such as *Rudbeckia, Kalmia, Solandra, Sigesbeckia,* and so forth, according to Wilfrid Blunt, *The Complete Naturalist: A Life of Linnaeus* (London: Frances Lincoln, 2001).

20. Johann Wolfgang von Goethe, "Formation and Transformation," in *Goethe's Botanical Writings,* trans. Bertha Mueller (Woodbridge, Conn.: Ox Bow Press, 1989), 23. Goethe's essay, although only a few pages in length, was written over a number of years in defense of the critique on his early work and was originally titled "Wandelns und umwandelns."

21. Arber was the first to translate Goethe's botanical scholarship, introducing German naturalism to the English-speaking world. Agnes Arber, *Goethe's Botany* (Waltham: Chronica Botanica Company, 1946).

22. The natural sciences advanced significantly at this time with the discovery of cellular units, substantiated by Robert Hooke (1635–1703) in *Micrographia* (1665) and Carl Linnaeus (1707–78), whose binomial taxonomy or *Systema Naturae* established the standard for classification when *Species Plantarum* was published and gained international prominence in 1753.

23. D. Von Mucke, "Goethe's Metamorphosis: Changing Forms in Nature, the Life Sciences, and Authorship," *Representations* 95, no. 1 (Summer 2006): 31.

24. Johann Wolfgang von Goethe, "Preliminary Notes for a Physiology of Plants," in *Goethe's Botanical Writings,* trans. Bertha Mueller (Woodbridge, Conn.: Ox Bow Press, 1989), 160.

25. I use the term *legislator* directly from James Larson's excellent account of the troubled relationship between the two naturalists. James L. Larson, "Goethe and Linnaeus," *Journal of the History of Ideas* 28, no. 4 (1967): 593.

26. Johann Wolfgang von Goethe, "The Experiment as Mediator between Object and Subject," in *Scientific Studies,* by Johann Wolfgang von Goethe, ed. and trans. Douglas Miller (Princeton, N.J.: Princeton University Press, 1995), 12.

27. Johann Wolfgang von Goethe, *Italian Journey,* trans. Robert R. Heitner, ed. Thomas P. Saine and Jeffrey L. Sammons (Princeton, N.J.: Princeton University Press, 1994).

28. Goethe, "The Experiment as Mediator," 11.

29. Goethe, *Italian Journey,* 310.

30. Johann Wolfgang von Goethe, "The Metamorphosis of Plants," in *Goethe's Botanical Writings,* trans. Bertha Mueller (Woodbridge, Conn.: Ox Bow Press, 1989).

31. For instance, arguments such as those by Donald R. Kaplan propose that morphology was not traditionally accepted in the United States, because it is "principally an engineering society, concerned more with the tools of science than with its theory, philosophy, and history." See Donald R. Kaplan, "The Science of Plant Morphology: Definition, History, and Role in Modern Biology," *American Journal of Botany* 88, no. 10 (2001): 1711–41.

32. Donald R. Kaplan, "The Teaching of Higher Plant Morphology in the United States," *Plant Science Bulletin* 19, no. 10 (1973): 6–9.

33. Arber defines plant morphology as the "comparative examination of form, studied in itself, and for its own sake." Agnes Arber, *Monocotyledons; a Morphological Study,* Cambridge Botanical Handbooks (Cambridge: Cambridge University Press, 1925), 1.

34. For an excellent short biography, see Rudolf Schmid, "Agnes Arber, Née Robertson (1879–1960): Fragments of Her Life, Including Her Place in Biology and in Women's Studies," *Annals of Botany* 88, no. 6 (2001): 1105–28.

35. Arber, *The Natural Philosophy of Plant Form,* 6–7. Arber's examination of the scientific method is most notable in Agnes Arber, *The Mind and the Eye: A Study of the Biologist's Standpoint* (Cambridge: Cambridge University Press, 1964).

36. Peter Bell, "'The Natural Philosophy of Plant Form.' By Agnes Arber (Book Review)," *British Journal for the Philosophy of Science* 1, no. 4 (1951): 336.

37. Few studies compare Goethe and Arber's artistic sensibilities. The following personal account is most aligned with the comparison: Maura C. Flannery, "Goethe & Arber: Unity in Diversity," *American Biology Teacher* 57, no. 8 (1995): 544–47.

38. Arber, *The Natural Philosophy of Plant Form,* 1.

39. Arber, *The Mind and the Eye,* 23.

40. Arber, 4.

41. The appearance of autotrophic and heterotrophic organisms continues to be debated. For clarity, see the pioneering work of Lynn Margulis: Lynn Margulis, *Early Life* (Boston: Science Books International, 1982).

42. Lynn Margulis and Dorian Sagan, "Life's Body," in *What Is Life?* (New York: Simon & Schuster, 1995), 4.

43. Joseph E. Armstrong, *How the Earth Turned Green: A Brief 3.8-Billion-Year History of Plants* (Chicago: University of Chicago Press, 2014), 33.

44. For an excellent and accessible description of the evolution of bacteria, see Armstrong, chapter 2, "Small Green Beginnings," 31–73.

45. Lynn Margulis, *Symbiotic Planet: A New Look at Evolution* (New York: Basic Books, 1998), 119. Margulis explains how Darwin laid the foundation for evolutionary biology and shows that symbiosis is the most crucial evolutionary novelty. Evolution was not a term used by Darwin (1802–89); rather, he describes "descent with modification" in his early treatise as a relentless display of mutations and refinements in all scales of living matter.

46. Lynn Margulis and Dorion Sagan, *Acquiring Genomes: A Theory of the Origins of Species* (New York: Basic Books, 2002), 12.

47. Margulis and Sagan, *What Is Life?,* 177.

48. Armstrong, *How the Earth Turned Green,* 59–63.

49. Armstrong, 188–90. This is of course much more nuanced, but Armstrong provides a few interesting pages of review on the land plant cycle, which concludes: "To enhance dispersal, the sporophyte needs to be as tall as possible, and so sporangia are placed terminally on the tall upright axis or on a short axis that is held aloft by a branch of the gametophyte."

50. Margulis, *Symbiotic Planet*, 107.

2. GREAT GREEN WALL

1. David O'Connor and James Ford, "Increasing the Effectiveness of the 'Great Green Wall' as an Adaptation to the Effects of Climate Change and Desertification in the Sahel," *Sustainability* 6, no. 10 (2014): 7142–54; Agence Panafricaine de la Grande Muraille Verte, "Approche conceptuelle," in *Strategie 2016–2020: Document Cadre*, 27–28, accessed October 21, 2021, www.grandemurailleverte.org.

2. "Harmonised Regional Strategy for Implementation of the 'Great Green Wall Initiative of the Sahara and the Sahel,'" New Partnership for Africa's Development (NEPAD), September 7, 2020, http://www.fao.org/fileadmin/templates/europeanunion/pdf/harmonized_strategy_GGWSSI-EN.pdf.

3. In reference to the Great Green Wall, Legasse Negash, a plant physiologist at Addis Ababa University, points out: "African leaders should do better to develop smart policies (and implement these within their means), instead of proposing pompous ideas owned by nobody." Legasse Negash, email to author, April 2020.

4. According to NEPAD, the GGW project expanded over a decade to include integrated ecosystem management, although the language of a "wall" initiated support through project development. See Emeka Johnkingsley, "'Green Wall' Project Gathers Pace in Senegal," August 12, 2011, http://www.SciDev.Net.

5. "Ce sont de vrais déserts que naissent aujourd'hui, sous nos yeux, dans des pays où il tombe cependant annuellement de 700 à plus de 1,500 mm de pluies." André Aubréville, *Climats, forêts et désertification de l'Afrique tropicale* (Paris: Soc. d'éd. géographiques, maritimes et coloniales, 1949), 332. Translations are mine unless otherwise indicated.

6. "Il n'est pas nécessaire d'être un visionnaire pour appercevoir sans erreur possible l'image de l'Afrique future. . . . L'Afrique tend sur l'échelle biologiques, vers la savane: savane nue." Aubréville, 329.

7. Aubréville, 211.

8. Notably, Hubert was trained as a naturalist and was chief administrator for AOF in 1924. H. Hubert, *Le dessèchement progressif en Afrique Occidentale* (Paris: Bulletin de Comité d'Etudes Historiques et Scientifiques d'AOF, 1920), 401–67.

9. E. William Bovill, "The Encroachment of the Sahara on the Sudan," *Journal of the Royal African Society* 20, no. 79 (1921): 174–85.

10. See especially the paper presented at the Royal Geographic Society with the Institute of British Geographers: Edward Percy Stebbing, "The Encroaching Sahara: The Threat to the West African Colonies," in *Geographical Journal* 85, no.6 (1935): 506–19, expanded upon in Edward Percy Stebbing, *The Forests of West Africa and the Sahara: A Study of Modern Conditions* (London: W. & R. Chambers, 1937).

11. See Diana K. Davis, "Imperialism and the Desert Blame Game," in *The Arid Lands: History, Power, Knowledge* (Cambridge, Mass.: MIT Press, 2016), 81–116.

12. Aubréville was greatly influenced by his particular education as an engineer at L'École Nationale des Eaux et des Forêts (ENEF, French National School of Forestry), which was one of the first in the world to advance rural forestry and silviculture. The school flourished under the directorship of Philibert Guinier, who is sometimes referred to as the father of forest ecology. The school also prepared the most influential

colonial scientists who facilitated the sequestration of natural resources from North Africa. Aubréville was a specialist in tropical botany, complemented by a high level of expertise in hand rendering and plant identification.

13. Helmut Geist, *The Causes and Progression of Desertification* (London: Aldershot, 2005), 12.

14. United Nations Convention to Combat Desertification (UNCCD), *Down to Earth: A Simplified Guide to the Convention to Combat Desertification* (Bonn, Germany: Secretariat for the United Nations Convention to Combat Desertification, 1995), 4.

15. Paula J. Williams, "Traditional Agroforestry," November 1, 1984, http://www.icwa.org/wp-content/uploads/2015/10/PJW-12.pdf.

16. For an overview of forest policy, see J. M. Boffa, "Institutional Factors in Parkland Management," in *Agroforestry Parklands in Sub-Saharan Africa* (Rome: FAO, 1999), 115–54.

17. "FAO's State of the World's Forests 2014," *Forestry Chronicle* 90, no. 5 (2014): 564.

18. See, for instance, Roni Avissar and David Werth, "Global Hydroclimatological Teleconnections Resulting from Tropical Deforestation," *Journal of Hydrometeorology* 6, no. 2 (2005): 134–45; Jule Charney et al., "A Comparative Study of the Effects of Albedo Change on Drought in Semi-Arid Regions," *Journal of the Atmospheric Sciences* 34, no. 9 (1977): 1366–85; and David Werth and Roni Avissar, "The Local and Global Effects of African Deforestation," *Geophysical Research Letters* 32, no. 12 (2005): L12704.

19. The work of Sam Moyo in Zimbabwe is relevant to this argument. Sam Moyo, *African Land Questions, Agrarian Transitions and the State: Contradictions of Neo-liberal Land Reforms* (Dakar, Senegal: Council for the Development of Social Science Research in Africa, 2008).

20. Significant literature on the tradition exists; see, for instance, Jesse Ribot, "Decentralisation, Participation and Accountability in Sahelian Forestry: Legal Instruments of Political-Administrative Control," *Africa (pre-2011)* 69, no. 1 (1999): 23–65.

21. Forestry code was ratified on July 4, 1935, transferring control of resources to French West Africa (Mauritania, Senegal, Guinea, Sudan, Niger). GGAOF (Gouvernement Générale de l'Afrique Occidentale Française), "No. 1704 A.P.—Arrêté prémulgant en Afrique occidentale française le décret du 4 juillet 1935, sur le régime forestier de l' Afrique occidentale française," *Journal Officiel du Sénégal,* July 24, 1935, 599–606.

22. Moyo, *African Land Questions*, 6.

23. "Africa's Great Green Wall Reaches Out to New Partners," FAO, December 16, 2013.

24. "Living on Earth: Africa's Great Green Wall of Trees," *Living on Earth,* March 30, 2012.

25. Mark Hertsgaard, "A Great Green Wall for Africa?," *The Nation* 293, no. 21 (2011): 22

26. For an excellent firsthand report of many "pilot" projects, see Jori Lewis, "The Forest for the Trees," *The Crisis* 121, no. 1 (2014): 22–27.

27. See, for instance, Johannes Schuler, Anna Katharina Voss, Hycenth Tim Ndah, Karim Traore, and Jan De Graaff, "A Socioeconomic Analysis of the Zaï Farming Practice in Northern Burkina Faso," *Agroecology and Sustainable Food Systems* 40, no. 9 (2016): 988–1007; and Robert Zougmoré, Abdulai Jalloh, and Andre Tioro, "Climate-Smart Soil Water and Nutrient Management Options in Semiarid West Africa: A Review of Evidence and Analysis of Stone Bunds and *Zaï* Techniques," *Agriculture & Food Security* 3, no. 1 (2014).

28. The work of Chris Reij is especially relevant to decentralized management. See Chris Reij and Robert Winterbottom, *Scaling Up Regreening: Six Steps to Success* (Washington, D.C.: World Resources Institute, 2015).

29. For project specifics, see "REAAP," Catholic Relief Services, accessed January 2020, https://www.crs.org/our-work-overseas/program-areas/agriculture/reaap.

30. Anthony Trewavas maintains that a record of past environments is deposited and stored in protein phosphorylation states, an activity of cell proliferation. Anthony Trewavas, "Intelligence and Consciousness," in *Plant Behaviour and Intelligence* (Oxford: Oxford University Press, 2014).

31. In his survey of tree planting in the sub-Sahel, Gerald Leach emphasizes three points: estimates of tree stocks and tree resources are rough in many cases; these estimates are held by forest departments, which know little about the volumes of trees outside the "forest," for example on village commons; and Gap predictions assume that once a hectare of forest has been cut, it is "dead land." Gerald Leach and Robin Mearns, *Beyond the Woodfuel Crisis: People, Land, and Trees in Africa* (London: Earthscan Publications, 1988), 8.

32. F. White, Association pour l'étude taxonomique de la flore d'Afrique tropicale, Comité pour la Carte de végétation, Oxford University Press, and UNESCO, *UNESCO/AETFAT/UNSO Vegetation Map of Africa* (Paris: UNESCO, 1981).

33. F. White, *The Vegetation of Africa: A Descriptive Memoir to Accompany the UNESCO/AETFAT/UNSO Vegetation Map of Africa* (3 plates: Northwestern Africa, Northeastern Africa, and Southern Africa, 1:5,000,000) (Paris: UNESCO, 1983).

34. The drought of 1968–73 led to extensive research into historical variations in rainfall in the Sahel. Conclusions drawn were that the level of drought (often termed desertification) fell well within the range of variability of rainfalls experienced over the last few centuries. See Sharon Nicholson, "The Climatology of Sub-Saharan Africa," in *Environmental Change in the West African Sahel* (Washington, D.C.: National Academy Press, 1982), 71–90.

35. An influential publication brought the crisis to Africa: "For more than a third of the world's people, the real energy crisis is a daily scramble to find the wood they need to cook dinner." Erik P. Eckholm, *The Other Energy Crisis: Firewood* (Washington, D.C.: Worldwatch Institute, 1975), 1.

36. Paula J. Williams, "(No Longer) Blowin' in the Wind," Institute of Current World Affairs, February 28, 1985, www.icwa.org/wp-content/uploads/2015/10/PJW-15.pdf.

37. The project was funded by the Cooperative for Assistance and Relief Everywhere (CARE), the lead agency for the project, and the subsequent development of agriculture and natural resources (ANR) programs. See, for instance, P. F. Ffolliott and R. L. Jemison, "Land Use in the Majjia Valley, Niger, West Africa," *General Technical Report RM—Rocky Mountain Forest and Range Experiment Station, United States, Forest Service,* no. 120 (1985): 470–74; Michael Angstreich, "Are Trees Effective against Desertification? Experiences from Niger and Mali," in *When the Grass Is Gone: Development Intervention in African Arid Lands* (Uppsala: Scandinavian Institute of African Studies, 1991), 141–51; and Leach and Mearns, *Beyond the Woodfuel Crisis.*

38. J. Sumberg and M. Burke, "People, Trees and Projects: A Review of CARE's Activities in West Africa," *Agroforestry Systems* 15, no. 1 (1991): 70.

39. A firsthand description of the specific environmental and cultural context from the time is covered in Angstreich, "Are Trees Effective against Desertification?," 142.

40. Leach and Mearns, *Beyond the Woodfuel Crisis,* 39.

41. Williams, "(No Longer) Blowin' in the Wind," 3.

42. Williams, 3.

43. Angstreich, "Are Trees Effective against Desertification?," 143.

44. Leach and Mearns, *Beyond the Woodfuel Crisis,* 154.

45. W. M. Ciesla, "What Is Happening to the Neem in the Sahel," *Unasylva,* no. 172 (1993): 45–51.

46. Leach and Mearns, *Beyond the Woodfuel Crisis,* 154.

47. "Eucalyptus camaldulensis," Agroforestry Database, accessed April 5, 2020, http://www.worldagro forestry.org/treedb/AFTPDFS/Eucalyptus_camaldulensis.PDF.

48. Williams, "(No Longer) Blowin' in the Wind," 5. Williams, in a direct account from the field in 1985, explains: "The sociological survey, supervised by James Delehanty, Marilyn Hoskins, and James Thomson, was conducted between May and July 1984. Six local Hausa-speaking interviewers, three women and three men, questioned 211 local women and 209 local men on the project's desirability and on other agroforestry practices."

49. Kathleen Buckingham and Craig Hanson, *The Restoration Diagnostic, Case Example: Maradi and Zinder Regions, Niger* (Washington, D.C.: World Resources Institute, 2015), 5.

3. Genus *Faidherbia*

1. O. Roupsard et al., "Reverse Phenology and Dry-Season Water Uptake by *Faidherbia albida* (Del.) A. Chev. in an Agroforestry Parkland of Sudanese West Africa," *Functional Ecology* 13, no. 4 (1999): 460–72.

2. For more on why *Faidherbia* is categorically uncommon, see P. J. Wood, "Botany and Distribution of *Faidherbia albida,*" in *Faidherbia Albida in the West African Semi-arid Tropics: Proceedings of a Workshop, 22–26 Apr 1991, Niamey, Niger,* ed. R. J. Vandenbeldt (Nairobi, Kenya: International Centre for Research in Agroforestry, 1992), 9–17.

3. "*Acacieae* Dumort 1829," in *Legumes of the World,* ed. Gwilym P. Lewis et al. (Kew: Royal Botanic Gardens, 2005), 187.

4. For instance, in an experiment of self-selection from Kew Gardens (UK) at Zalingei, Sudan, high seedling mortality rates were recorded at 96 percent. G. E. Wickens, "A Study of *Acacia albida* Del. (*Mimosoïdeae*)," *Kew Bulletin* 23, no. 2 (1969): 194.

5. P. J. Dart, "Microbial Symbioses of Tree and Shrub Legumes," in *Forage Tree Legumes in Tropical Agriculture,* ed. Ross Gutteridge and H. Max Shelton (Wallingford, Oxford: CAB International, 1994), 145.

6. Roupsard et al., "Reverse Phenology," 471.

7. Y. Dalpé et al., "Glomales Species Associated with Surface and Deep Rhizosphere of *Faidherbia albida* in Senegal," *Mycorrhiza* 10, no. 3 (2000): 125–29; Stone and Kalisz, "On the Maximum Extent of Tree Roots," *Forest Ecology and Management* 46, no. 1 (1991): 59–102; Nicolas C. Dupuy and Bernard L. Dreyfus, "*Bradyrhizobium* Populations Occur in Deep Soil under the Leguminous Tree *Acacia albida,*" *Applied and Environmental Microbiology* 58, no. 8 (1992): 2415–19.

8. See, for instance, Wickens, "A Study of *Acacia albida* Del. (*Mimosoïdeae*)," 181–202.

9. J. E. Beringer et al., "The Rhizobium-Legume Symbiosis," *Proceedings of the Royal Society of London, Series B, Biological Sciences (1934–1990)* 204, no. 1155 (1979): 219–33.

10. Agnes Arber, *The Gramineae: A Study of Cereal, Bamboo, and Grass* (Cambridge: Cambridge University Press, 1934), 248.

11. Dart, "Microbial Symbioses of Tree and Shrub Legumes," 146. For more on nodulation processes, see Janet I. Sprent, *Nodulation in Legumes* (Kew: Royal Botanic Gardens, 2001).

12. See, for instance, Janet I. Sprent and Richard Parsons, "Nitrogen Fixation in Legume and Non-legume Trees," *Field Crops Research* 65, no. 2 (2000): 183–96. Dart comments on the plant-derived membrane: "The rhizobia invade the root growing in the intercellular spaces, a microcolony causes the cortical

cell wall to thin and grow around the bacteria, depositing cell wall materials on the internal wall surface, and the bacteria appear to grow through this barrier. Once released, the rhizobia induce host cell division and divide along with the chromosomes in the process." Dart, "Microbial Symbioses of Tree and Shrub Legumes," 145.

13. E. L. Simms, "Partner Choice in Nitrogen-Fixation Mutualisms of Legumes and Rhizobia," *Integrative and Comparative Biology* 42, no. 2 (2002): 369–80.

14. Such debates are typically framed through overgrazing, desertification, or resource depletion. See H. N. Le Houérou, *The Grazing Land Ecosystems of the African Sahel,* Ecological Studies 75 (Berlin: Springer-Verlag, 1989), 63.

15. Joseph Armstrong in particular argues that there is a "modern vegetation" that developed through exclusive human interaction such as agriculture. See Joseph Armstrong, *How the Earth Turned Green: A Brief 3.8-Billion-Year History of Plants* (Chicago: University of Chicago Press, 2014), chapter 11.

16. Armstrong, 348.

17. At present, 11 percent of the globe's land surface is used for crop production. For this reason, the Anthropocene is also termed the plantationocene by Haraway and others. See Donna Haraway, "Anthropocene, Capitalocene, Plantationocene, Chthulucene: Making Kin," *Environmental Humanities* 6 (2015): 159–65.

18. An excellent account of banding and scatter patterns is found in D. Tongway, C. Valentin, and J. Seghieri, *Banded Vegetation Patterning in Arid and Semi-arid Environments: Ecological Processes and Consequences for Management,* Ecological Studies 149 (New York: Springer, 2001).

19. Richard Karban, *Plant Sensing and Communication* (Chicago: University of Chicago Press, 2016), 34.

20. This is the basis of Darwin's concept of "natural selection," whereby all organisms produce more offspring than can survive. Particular to *Faidherbia,* see P. Hauser, "Germination, Predation and Dispersal of *Acacia albida* Seeds," *Oikos* 70, no. 3 (1994): 421–26.

21. This is a particular feature of flowering plants, or Angiosperms: "Prior to the appearance of flowering plants, the interaction between animals and plants could hardly be called 'cooperative'; animals fed upon plants, a rather one-sided interaction. Of course, animals, including ourselves, still feed upon plants, but flowering plants found a way to benefit from some of this animal feeding by using animals." Armstrong, *How the Earth Turned Green,* 3.

22. "It takes *Faidherbia albida* 15 years to have local benefits." Robert D. Kirmse and Brien E. Norton, "The Potential of *Acacia albida* for Desertification Control and Increased Productivity in Chad," *Biological Conservation* 29, no. 2 (1984): 137.

23. J. Ahée and E. Duhoux, "Root Culturing of Faidherbia: *Acacia albida* as a Source of Explants for Shoot Regeneration," *Plant Cell, Tissue and Organ Culture* 36, no. 2 (1994): 219–25.

24. Micropropogation is the use of biotechnology to grow a large number of plants from very small pieces of plants, including single cells, tissues, and seeds by in-vitro methods. Chris Park and Michael Allaby, "Micropropagation," in *A Dictionary of Environment and Conservation* (Oxford: Oxford University Press, 2013).

25. These associations are dependent on root hairs that emerge and elongate in a zone several millimeters behind the root tip, in most plants. See D.T. Clarkson, "Factors Affecting Mineral Nutrient Acquisition by Plants," *Annual Review of Plant Physiology* 36 (1985): 82; and Paola Bonfante and Iulia-Andra Anca, "Plants, Mycorrhizal Fungi, and Bacteria: A Network of Interactions," *Annual Review of Microbiology* 63 (2009): 363.

26. Much literature exists on mycorrhizal relationships, as their importance is coming to light in this century. For a general discussion, see Karban, *Plant Sensing and Communication*, 133–35. The literature is saturated with studies to improve growing conditions for *Faidherbia* with mycorrhizae. In Senegal, for instance, see A. T. Diallo, P. Samb, and M. Ducousso, "Arbuscular Mycorrhizal Fungi in the Semi-arid Areas of Senegal," *European Journal of Soil Biology* 35, no. 2 (1999): 65–75; F. Bernatchez et al., "Soil Fertility and Arbuscular Mycorrhizal Fungi Related to Trees Growing on Smallholder Farms in Senegal," *Journal of Arid Environments* 72 (2008): 1247–56; and Dalpé et al., "Glomales Species."

27. Bernatchez et al., "Soil Fertility," 1247.

28. Dart, "Microbial Symbioses of Tree and Shrub Legumes," 147. In this section of the report funded by FAO, the author cites an experiment with eucalyptus in Queensland, Australia, remarking that soils with low levels of mycorrhizal activity "were likely to have been continuously cultivated."

29. Jules-Emile Planchon, *The Eucalyptus Globulus from a Botanic, Economic, and Medical Point of View, Embracing Its Introduction, Culture, and Uses,* Bulletin of the U.S. Department of Agriculture 9 (Washington, D.C.: Government Printing Office, 1875), 14.

30. For an interesting account of Francois Trottier and the failure of early eucalyptus plantations in Algiers, see Diana K. Davis, *Resurrecting the Granary of Rome: Environmental History and French Colonial Expansion in North Africa* (Athens: Ohio University Press, 2007), 105–8.

Index

1. Gifford Pinchot, *The Profession of Forestry* (Washington, D.C.: American Forestry Association, 1901), 1.

4. Confronting Treelessness

1. Reuel R. Hanks and Stephen John Stadler, "Grasslands," in *Encyclopedia of Geography Terms, Themes, and Concepts* (Santa Barbara, Calif.: ABC-CLIO, 2011), 160–62.

2. For an excellent description of the similarity in flora and fauna between European and North American ecology, as evidenced until the nineteenth century, see Alfred W. Crosby, *Ecological Imperialism: The Biological Expansion of Europe, 900–1900* (Cambridge: Cambridge University Press, 1986). The expansion of horticulture and especially fruit trees in America after the nineteenth century is chronicled in Philip J. Pauly, *Fruits and Plains: The Horticultural Transformation of America* (Cambridge, Mass.: Harvard University Press, 2007).

3. S. E. Cohen, *Planting Nature: Trees and the Manipulation of Environmental Stewardship in America* (Berkeley: University of California Press, 2004), 26–30.

4. Herbert Quick, *Vandermark's Folly* (New York: Curtis, 1916), 229. *Vandermark's Folly* was a popular novel at the time; it describes the highway of pastoral immigrants especially though Wisconsin and Iowa into Nebraska.

5. The United States Department of Agriculture (USDA) was founded in 1862, signed into law by an act of congress. For more, see, for instance, Carleton Roy Ball, *History of the U.S. Department of Agriculture and the Development of Its Objectives* (Washington, D.C.: Dept. of Agriculture, 1936).

6. For a description of the life cycles common to describing the prairie as an organism, see F. E. Clements, *Plant Succession: An Analysis of the Development of Vegetation* (New York: Wilson Company, 1928).

7. Richard Manning, *Grassland: The History, Biology, Politics, and Promise of the American Prairie* (New York: Viking, 1995), 71.

8. Manning, 58.

9. The full Timber Culture Act is available in Thomas Donaldson, *The Public Domain: Its History, with Statistics* (Washington, D.C.: Government Printing Office, 1884), 1093.

10. Roxanne Dunbar-Ortiz, *An Indigenous Peoples' History of the United States* (Boston: Beacon, 2014).

11. For more on Nebraska as a "tree-planting state," see John F. Freeman, "Trees for High Plains," in *High Plains Horticulture: A History* (Boulder: University Press of Colorado, 2008), 19–32.

12. Donaldson, *The Public Domain*, 1092.

13. Donaldson, 1094; Timber Culture Act, application October 3, 1878, Individual Claimant, NB.

14. Benjamin H. Hibbard, *A History of the Public Land Policies* (Madison: University of Wisconsin Press, 1965), 422.

15. Accounts of individual claims are covered in relation to the landscape in Charles Barron McIntosh, *The Nebraska Sand Hills: The Human Landscape* (Lincoln: University of Nebraska Press, 1996).

16. Everett Newfon Dick, *Conquering the Great American Desert: Nebraska* (Lincoln: Nebraska State Historical Society, 1975).

17. Philip J. Pauly, *Fruits and Plains: The Horticultural Transformation of America* (Cambridge, Mass.: Harvard University Press, 2007).

18. C. B. McIntosh, "Use and Abuse of the *Timber Culture Act*," *Annals of the Association of American Geographers* 65, no. 3 (1975): 347–62.

19. In a description of the region, an introductory study by Raphael Zon of the Forest Service explains that "the climatic conditions become less favorable for plant growth from east to west." Lake States Forest Experiment Station, U.S. Forest Service, *Possibilities of Shelterbelt Planting in the Plains Region* (Washington, D.C.: Government Printing Office, 1935), 3.

20. Lake States Forest Experiment Station, 7.

21. Charles Sprague Sargent, *The Silva of North America: A Description of the Trees Which Grow Naturally in North America Exclusive of Mexico* (Boston: Houghton, Mifflin, 1891), v.

22. See also Charles Sprague Sargent, *Sixteen Maps Accompanying Report on Forest Trees of North America* (Washington, D.C.: U.S. Census Office, 1884).

23. Sargent, *The Silva of North America*, v.

24. The term *ecology* was first proposed by German biologist Ernst Haeckel (1834–1919), after reviewing Darwin's *On the Origin of Species* (1834), but was not widely accepted. I am pointing out that its most formative adoption came through the plant sciences and was applied and practiced by American grassland botanists. At the time, formative texts on ecology were being published by professors of botany, notably Eugenius Warming, *Oecology of Plants; an Introduction to the Study of Plant-Communities* (Oxford: Clarendon, 1909); and Frederic Edward Clements, *Research Methods in Ecology* (Lincoln, Neb.: University Publishing Company, 1905).

25. John E. Weaver and Frederic E. Clements, *Plant Ecology*, 2nd ed. (New York: McGraw-Hill, 1938). For an excellent account of the botanical influence on ecology in America, see also Sharon E. Kingsland, *The Evolution of American Ecology* (Baltimore: Johns Hopkins University Press, 2005); and Paul B. Sears, "Plant Ecology," in *A Short History of Botany in the United States*, ed. Joseph Ewan (New York: Hafner, 1969), 124–31.

26. On the origins of plant ecology, see Ronald C. Tobey, *Saving the Prairies: The Life Cycle of the Founding School of American Plant Ecology, 1895–1955* (Berkeley: University of California Press, 1981).

27. John Ernest Weaver, "A Study of the Root-Systems of Prairie Plants of Southeastern Washington," *Plant World* 18, no. 9 (1915): 227–48.

28. Rhizography is a term used to describe the distribution of roots; it first appeared in a 1975 paper by W. H. Lyford titled "Rhizography of Non-woody Roots of Trees in the Forest Floor." Lyford's research broadened the scope of study toward the living environment by including the analysis of the relationship between roots. By relating the study of roots to the discipline of geography, he contributed to a broader understanding of plant life within the physical features of the earth.

29. Weaver published hundreds of works on plant botany, grassland ecology, and root morphology. In particular, see *The Ecological Relation of Roots* (Washington, D.C.: Carnegie Institution of Washington, 1919), "Investigations on the Root Habits of Plants," *American Journal of Botany* 12, no. 8 (1925): 502–9; "Root Distribution of Trees in Relation to Soil Profile," *Ecology* 19, no. 1 (1938): 156–57; *North American Prairie* (Lincoln: University of Nebraska, 1944), "The Living Network in Prairie Soil," *Botanical Gazette* 123, no. 1 (1961): 16–28. His complete papers are available through the University of Nebraska, "Papers of John E. Weaver (1884-1956)," Digital Commons, accessed September 22, 2021, http://digitalcommons.unl.edu/agronweaver/.

30. John E. Weaver, *Prairie Plants and Their Environment* (Lincoln: University of Nebraska Press, 1968), 51. Here Weaver refers to plant species, not the hundreds of thousands of other species that populate the soil horizons.

31. John E. Weaver, "Investigations on the Root Habits of Plants," *American Journal of Botany* 12, no. 8 (1925): 502.

32. Weaver, *Prairie Plants and Their Environment,* 2: "The individual root habit and especially the community root habit, together with the more familiar above-ground parts, serve to interpret the environmental conditions."

33. The lineage of plant ecology can be traced through the Department of Botany at the University of Nebraska, where the head of the department, botanist Charles Bessey, trained Clements. Shortly after Bessey's death, Weaver joined the department, where he remained for his entire career (1917–52), whereas Clements moved on to various positions, mainly the pursuit of ecological scholarship. Clements coined the terms *climax* and *succession* to describe disturbance but only saw it as a mechanism to reach equilibrium. Nonetheless, advances at the University of Nebraska are especially critical to how the term *ecology* emerged from botany.

34. Frederic E. Clements, *Research Methods in Ecology* (Lincoln, Neb.: Jacob North, 1905), 199.

35. Weaver and Clements, *Plant Ecology.*

36. David J. Wishart, *Great Plains Indians* (Lincoln, Neb.: Bison Books, 2016), 5.

37. George P. Marsh, *Man and Nature* (New York: Charles Scribner, 1864), 14.

38. On the embrace of Pinchot's policies by Americans, see Michael Williams, *Americans and Their Forests: A Historical Geography* (Cambridge: Cambridge University Press, 1989), 393–411.

39. Gifford Pinchot, *The Fight for Conservation* (New York: Doubleday, Page, 1910).

40. Gifford Pinchot, *A Primer of Forestry* (Washington: Government Printing Office, 1903), 7.

41. Pinchot, *The Fight for Conservation,* iv.

42. David Lowenthal, *George Perkins Marsh, Prophet of Conservation,* Weyerhaeuser Environmental Book (Seattle: University of Washington Press, 2000), 9.

43. Lowenthal, 10–12.

44. Office of Information, U.S. Forest Service, *Prairie-Plains Region Shelterbelt Project* (Washington, D.C.: U.S. Forest Service, 1946).

45. Wilmon H. Droze, *Trees, Prairies, and People* (Denton: Texas Woman's University, 1977), 30.

46. Droze, 22.

47. There is tremendous literature on the history of conservation. See, for instance, George Perkins Marsh, *Man and Nature,* ed. David Lowenthal (Seattle: University of Washington Press, 1864); and Douglas Helms and Susan L. Flader, eds., *The History of Soil and Water Conservation* (Washington, D.C.: Agricultural Society, 1985).

5. PRAIRIE STATES FORESTRY PROJECT

1. Lake States Forest Experiment Station, U.S. Forest Service, *Possibilities of Shelterbelt Planting in the Plains Region* (Washington, D.C.: Government Printing Office, 1935), 1.

2. Neil Smith, "There's No Such Thing as a Natural Disaster," Items, June 11, 2006. https://items.ssrc .org. This seminal work of social sciences was published online following Hurricane Katrina.

3. Lake States Forest Experiment Station, *Possibilities of Shelterbelt Planting,* 3.

4. See U.S. Forest Service, "Land Areas of the National Forest System," September 30, 2012, https:// www.fs.fed.us/land/staff/lar/LAR2012/LAR_Book_FY2012_A4.pdf.

5. In particular, the Timber Culture Act confirmed landholdings. See C. Barron McIntosh, "Use and Abuse of the Timber Culture Act," *Annals of the Association of American Geographers* 65, no. 3 (1975): 347–62.

6. For an academic overview of settlement, see Ronald C. Naugle, John J. Montag, and James C. Olson, *History of Nebraska* (Lincoln: University of Nebraska Press, 2015). For a particularly powerful personal narrative of dispossession, see Susan Bordeaux Bettelyoun, Josephine Waggoner, and Emily Levine, *With My Own Eyes: A Lakota Woman Tells Her People's History* (Lincoln: University of Nebraska Press, 1998).

7. Charles Barron McIntosh, *The Nebraska Sand Hills* (Lincoln: University of Nebraska Press, 1996), 1.

8. For many fascinating accounts of life among the Pawnee, see David J. Wishart, *Great Plains Indians* (Lincoln, Neb.: Bison Books, 2016).

9. Bordeaux Bettelyoun, Waggoner, and Levine, *With My Own Eyes,* 86.

10. McIntosh, *The Nebraska Sand Hills,* 111.

11. An overview of the process is provided by an early history: Nathaniel Hillyer Egleston, *Arbor Day: Its History and Observance* (Washington, D.C.: Government Printing Office, 1896).

12. For a description of landscape features and the establishment of national grasslands, see Francis Moul, *The National Grasslands: A Guide to America's Undiscovered Treasures* (Lincoln: University of Nebraska Press, 2006).

13. Axel Rydberg, *Flora of the Sand Hills of Nebraska* (Washington, D.C.: Government Printing Office, 1895), 135.

14. Rydberg, 137.

15. Weaver's grassland experiments are particularly relevant. See, for instance, John E. Weaver and E. Zink, "Length of Life of Roots of Ten Species of Perennial Range and Pasture Grasses," *Plant Physiology* 21, no. 2 (1946): 201–17. See also Raymond J. Pool, "A Study of the Vegetation of the Sandhills of Nebraska" (PhD diss., University of Nebraska, 1914).

16. Entries of patents are taken from maps produced by Charles Barron McIntosh. See, for instance, C. Barron McIntosh, "Patterns from Land Alienation Maps," *Annals of the Association of American Geographers* 66, no. 4 (1976): 570–82.

17. The Bessey Nursery was developed by the Bureau of Forestry, later the Forest Service of the U.S. Department of Agriculture. Clark Fleege, "History of Bessey Nursery," in *National Proceedings: Forest and*

Conservation Nursery Associations, by T. D. Landis and B. Cregg, tech. coords. (Portland, Ore.: U.S. Department of Agriculture, 1995), 60–63.

18. John F. Freeman, *High Plains Horticulture: A History* (Boulder: University Press of Colorado, 2008), 72–74.

19. Charles Anderson Scott, *The Early Days: The Dismal River & Niobrara Forest Reserves* (Washington, D.C.: USDA Forest Service, 2002), 27, 33.

20. W. H. Droze, "Changing the Plains Environment: The Afforestation of the Trans-Mississippi West," *Agricultural History* 51, no. 1 (1977): 13.

21. Scott, *The Early Days,* 35.

22. *Maclura* was used primarily for its wood by indigenous tribes and as living fences by expansionists. For an explanation of possible endemic boundaries, see Jeffrey L. Smith and Janice V. Perino, "Osage Orange (*Maclura pomifera*): History and Economic Uses," *Economic Botany* 35, no. 1 (1981): 24–41.

23. John J. Winberry, "The Osage Orange, a Botanical Artifact," *Pioneer America* 11, no. 3 (1979): 134–41.

24. See USDA, "Osage Orange Plant Fact Sheet," accessed June 30, 2019, https://plants.usda.gov/fact sheet/pdf/fs_mapo.pdf.

25. Scott, *The Early Days,* 18.

26. Actual numbers range depending on the report; 125,000 seems to be most common. See Droze, "Changing the Plains Environment," 18.

27. Scott, *The Early Days,* 34.

28. Peter Del Tredici, "Saving Nature in a Humanized World: Setting the Context" (lecture, Cultural Landscape Foundation, San Francisco, Calif., January 23, 2015), https://www.youtube.com/watch?v=NX NEUkHHzgc.

29. See, for instance, Piers M. Blaikie, *The Political Economy of Soil Erosion in Developing Countries,* Longman Development Studies (New York: Longman, 1985), 45; and Recognizing the Duty of the Federal Government to Create a Green New Deal, H. Res. 109, 116th Congress, 1st Session, February 7, 2019, https://www.congress.gov/116/bills/hres109/BILLS-116hres109ih.pdf.

30. Historian Donald Worster describes the origins of the term *dust bowl* and the ecology of the 1930s. See, for instance, Donald Worster, *Dust Bowl: The Southern Plains in the 1930s* (Oxford: Oxford University Press, 1979), 28–29.

31. For a comprehensive synthesis of the role of the Civilian Conservation Corps on conservation and the landscape, see Neil Maher, *Nature's New Deal* (Oxford: Oxford University Press, 2008).

32. Maher, *Nature's New Deal,* 52.

33. USDA Forest Service, *A National Plan for American Forestry,* Senate document 12, 73rd Congress, 1st Session (Washington, D.C.: Government Printing Office, 1933), I, 1076.

34. McIntosh, *The Nebraska Sand Hills,* chapter 2.

35. McIntosh, 9.

36. Scott, *The Early Days,* 27.

37. Lake States Forest Experiment Station, *Possibilities of Shelterbelt Planting,* 39–41.

38. Lake States Forest Experiment Station, 54.

6. *Ulmus pumila* L.

1. For an excellent account of how seed-bearing plants evolved with humans, see Carolyn Fry, *Seeds: A Natural History* (Chicago: University of Chicago Press, 2016), 16.

2. For a case-based review of seed migration, see Jonathan D. Sauer, *Plant Migration: The Dynamics of Geographic Patterning in Seed Plant Species* (Berkeley: University of California Press, 1988), 140.

3. Anthony Trewavas, *Plant Behaviour and Intelligence* (Oxford: Oxford University Press, 2014), 161.

4. Katherine Vayda, Kathleen Donohue, and Gabriela Alejandra Auge, "Within- and Trans-generational Plasticity: Seed Germination Responses to Light Quantity and Quality," *AoB PLANTS* 10, no. 3 (May 2018).

5. D. H. Jennings, A. J. Trewavas, and Society for Experimental Biology, *Plasticity in Plants,* Symposia of the Society for Experimental Biology 40 (Cambridge: Published for the Society for Experimental Biology by the Company of Biologists Limited, 1986).

6. Trewavas, *Plant Behaviour and Intelligence,* 159.

7. Walter E. Webb, "A Report on *Ulmus Pumila* in the Great Plains Region of the United States," *Journal of Forestry* 46, no. 4 (April 1948): 274. The author of the report was a landscape engineer for the U.S. Forest Service.

8. Webb, 274.

9. Choimaa Dulamsuren, Markus Hauck, Suran Nyambayar, Dalaikhuu Osokhjargal, and Christoph Leuschner, "Establishment of *Ulmus Pumila* Seedlings on Steppe Slopes of the Northern Mongolian Mountain Taiga," *Acta Oecologica* 35, no. 5 (2009): 563.

10. Two regional studies in particular explore root dynamics as they relate to seedling development. This indicates that the whole plant is considered in the uncertainty of the experiment, by including root-to-shoot dynamics. Both research teams also test in the deserts with wild seed, as opposed to in controlled nurseries, in order to modify seed. In China, see Jiao Tang et al., "Seed Burial Depth and Soil Water Content Affect Seedling Emergence and Growth of *Ulmus Pumila* Var. Sabulosa in the Horqin Sandy Land," *Sustainability* 8, no. 1 (2016): 1–10. In Mongolia, see Dulamsuren et al., "Establishment of *Ulmus Pumila* Seedlings."

11. Dulamsuren et al., "Establishment of *Ulmus Pumila* Seedlings," 564.

12. J. Philip Grime, *Plant Strategies, Vegetation Processes, and Ecosystem Properties,* 2nd ed. (New York: Wiley, 2001), 173.

13. Stress-tolerant plasticity is extensively covered in Grime, chapters 1 and 2.

14. K. Wesche, M. Pietsch, K. Ronnenberg, R. Undrakh, and I. Hensen, "Germination of Fresh and Frost-Treated Seeds from Dry Central Asian Steppes," *Seed Science Research* 16, no. 2 (2006): 123–36.

15. This study proposes ultimate burial at 0.5–1.0 centimeters. See Tang et al., "Seed Burial Depth," 1.

16. Tang et al., "Seed Burial Depth," 6–7.

17. Trewavas, *Plant Behaviour and Intelligence,* 160.

18. Michael Evenari, "Seed Physiology: Its History from Antiquity to the Beginning of the 20th Century," *Botanical Review* 50, no. 2 (1984): 130–31.

19. My description of seeds is inspired by Darwin's *The Power of Movements in Plants,* in which he describes how a series of isolated movements express a universal movement—when taken together. He proves that every part of every plant is continually in motion, highlighting that these movements occur along varying scales, especially the very smallest. The adjustments of the seed are delicately described as being analogous to a man thrown on his back and the movements to "right" oneself on all fours. Darwin's experiments are significant for his equal consideration of the root and shoot, recognizing the entire living plant as his collaborator. Charles Darwin, *The Power of Movement in Plants* (London: John Murray, 1880), 24–25, 106–8.

20. Dicot stands for dicotyledon, and thus contains two cotyledons, while monocots have one. Monocots include true grasses, orchids, and bamboos.

21. Anthony Trewavas, "Intelligence, Cognition, and Language of Green Plants," *Frontiers in Psychology* 7 (2016): 588.

22. For example, one specialist commented on *Ulmus pumila*: "Very favorable reports have been received from practically every section of the country." J. Burtt Davy, "Elms for the Semi-arid Regions of the Empire," *Empire Forestry Journal* 8, no. 1 (1929): 103.

23. Davy, 103.

24. See plant lists by state, tables 3 to 7 in Lake States Forest Experiment Station, U.S. Forest Service, *Possibilities of Shelterbelt Planting in the Plains Region: A Study of Tree Planting for Protective and Ameliorative Purposes as Recently Begun in the Shelterbelt Zone of North and South Dakota, Nebraska, Kansas, Oklahoma, and Texas by the Forest Service; Together with Information as to Climate, Soils, and Other Conditions Affecting Land Use and Tree Growth in the Region* (Washington, D.C.: Government Printing Office, 1935), 18–22.

25. J. E. Weaver, "The Living Network in Prairie Soils," *Botanical Gazette*, no. 1 (1961): 18.

26. Paul C. Guilkey, *Silvical Characteristics of American Elm (Ulmus Americana)* (St. Paul, Minn.: Lake States Forest Experiment Station, Forest Service, U.S. Dept. of Agriculture, 1957) (see "limiting factors," p. 18).

27. Lake States Forest Experiment Station, *Possibilities of Shelterbelt Planting*, 24.

28. Richard Manning, *Grassland: The History, Biology, Politics, and Promise of the American Prairie* (New York: Viking, 1995), 170.

29. "Extraction or cleaning of the seed will be done at the centrally located seed stations." Lake States Forest Experiment Station, *Possibilities of Shelterbelt Planting*, 23.

30. See ongoing invasion control measures: USDA and NRCS, "The PLANTS Database," accessed March 20, 2018, https://www.plants.usda.gov/java/; USDA, "Field Guide for Managing Siberian Elm in the Southwest," accessed March 20, 2018, http://www.fs.usda.gov/Internet/FSE_DOCUMENTS/stelp rdb5410128.pdf. There are a number of sources for statistics of eradication, for instance Faith Thompson Campbell and Scott E. Schlarbaum, *Fading Forests: North American Trees and the Threat of Exotic Pests* (New York: Natural Resources Defense Council, 1994); and Heidi Hirsch et al., "Intra- and Interspecific Hybridization in Invasive Siberian Elm," *Biological Invasions* 19, no. 6 (2017): 1889–1904.

31. Susan Wieseler and Plant Conservation Alliance, *Siberian Elm Fact Sheet* (Washington, D.C.: Plant Conservation Alliance, 2005), 1.

32. See Michael Westphal et al., "The Link between International Trade and the Global Distribution of Invasive Alien Species," *Biological Invasions* 10, no. 4 (2008): 391–98.

33. Peter Del Tredici, "Preface to the Second Edition," in *Wild Urban Plants of the Northeast*, 2nd ed. (Ithaca, N.Y.: Cornell University Press, 2020), xiii–iv. A deep history of weeds offers a particularly useful look at the "Europeanization" of land: Alfred W. Crosby, *Ecological Imperialism: Biological Expansion of Europe, 900–1900* (Cambridge: Cambridge University Press, 1996), 145–70.

34. In the United States alone, the total annual economic cost for combatting "Invasive Alien Species" (IAS) hovers in the range of $120 billion. David Pimentel, Rodolfo Zuniga, and Doug Morrison, "Update on the Environmental and Economic Costs Associated with Alien-Invasive Species in the United States," *Ecological Economics* 52, no. 3 (2005): 273–88. For the red list of threatened species, see "Invasive Species," International Union for Conservation of Nature (IUCN), accessed April 3, 2019, https://www.iucn.org/theme/species/our-work/invasive-species.

35. See, for instance, Pimentel, Zuniga, and Morrison, "Environmental and Economic Costs."

36. *Amaranth*, an ancient crop originating in the Americas, is explored as a high-protein grain or as a leafy vegetable. Grain *amaranth* species are important in different parts of the world and at different

times for several thousand years. For a history and recipes, see Fernando Divina and Marlene Divina, *Foods of the Americas: Native Recipes and Traditions* (Berkeley, Calif.: Ten Speed Press, 2004).

37. Renita D. Young and Tom Polansek, "Minnesota Finds Source of Invasive Weed on Conservation Land," Reuters, April 6, 2017, https://www.reuters.com; Gil Gullickson, "Defeating Palmer Amaranth," *Successful Farming Magazine,* April 5, 2017, https://www.agriculture.com.

38. William Neuman and Andrew Pollack, "Farmers Cope with Roundup-Resistant Weeds," *New York Times,* May 3, 2010.

39. It was a Chinese pharmacopeia that first illustrated the potential harm of chemicals and their relation to human health in the treatise *Shennong Ben Cao Jing* (25–220 A.D.). More than 267 plants with pesticidal activity were described for their benefit or harm with a focus on the root. The development of allelopathy as a science is covered in Chang-Hung Chou, "Introduction to Allelopathy," in *Allelopathy: A Physiological Process with Ecological Implications,* ed. Manuel J. Reigosa, Nuria Pedrol, and Luís González (Dordrecht, Netherlands: Springer, 2006), 1–9.

40. Chou, "Introduction to Allelopathy," 1.

41. In one study, the authors conclude that leaf litter from *U. pumila* reduced the growth of understory species. The authors also claim that "this is the first report demonstrating the allelopathic effect of *Ulmus pumila* leaf litter," indicating the unstudied topic of buried communication. See M. Esther Pérez-Corona, De las Heras Paloma, and Beatriz R. Vazquez de Aldana, "Allelopathic Potential of Invasive *Ulmus pumila* on Understory Plant Species," *Allelopathy Journal* 32, no. 1 (2013): 101–12.

42. Chakravarthula Manoharachary and Krishina G. Mukerji, "Rhizosphere Biology—an Overview," in *Microbial Activity in the Rhizosphere,* ed. K. G. Mukerji et al. (New York: Springer, 2006), 10.

43. D. K. McE. Kevan, "The Soil Fauna—Its Nature and Biology," in *Ecology of Soil-Borne Plant Pathogens: Prelude to Biological Control,* ed. Kenneth F. Baker and W. C. Snyder (Berkeley: University of California Press, 1965), 33–50.

44. In the introduction to the proceedings from the International Symposium on Factors Determining the Behavior of Plant Pathogens in Soil, held at the University of California, Berkeley, in 1963, reference is made to defining the rhizosphere and as a comparison to a social environment: "Indeed we are dealing with a unique environment inhabited by a 'society' of microorganisms, a field of activity around a central point, which is another dictionary definition for sphere." Elroy A. Curl and Bryan Truelove, *The Rhizosphere* (Berlin: Springer-Verlag, 1986), 9.

45. Manning, *Grassland,* 142.

46. Weaver, "The Living Network in Prairie Soils," 33.

47. In *Grassland,* Manning provides a description of the American grasslands though the potential for conservation and restoration. In this, he guides the reader though different grasses and forbs, although the book does not include the terms of anthropogenic influence beyond agriculture and growth industries.

TRACE

1. Diana K. Davis, *The Arid Lands: History, Power, Knowledge* (Cambridge, Mass.: MIT Press, 2016), 8.

7. CONTEXTUAL INDIFFERENCE

1. "Deng Xiaoping General Assembly Speech," UN General Assembly, January 1, 1974, http://www.unmultimedia.org/classics/asset/C817/C817/.

2. Henry Kissinger, *On China* (New York: Penguin Books, 2011), 303.

3. David H. Shinn and J. Eisenman, *China and Africa: A Century of Engagement* (Philadelphia: University of Pennsylvania Press, 2012), 42–43.

4. Kissinger, *On China,* 362.

5. There are few English translations of the modification, but the following provides a strong overview: Sen Wang, G. Cornelis van Kooten, and Bill Wilson, "Mosaic of Reform: Forest Policy in Post-1978 China," *Forest Policy and Economics* 6, no. 1 (2004): 71–83.

6. Chunyang Li, Jarkko Koskela, and Olavi Luukkanen, "Protective Forest Systems in China: Current Status, Problems and Perspectives," *Ambio* 28, no. 4 (1999): 341–45.

7. The UNEP was initiated following the United Nations Conference on the Human Environment (Stockholm Conference). See Carol A. Pentsonk, "The Role of United Nations Environment Programme (UNEP) in the Development of International Environmental Law," *American University International Law Review* 5, no. 2 (1990): 351–91.

8. C. Raynaut et al., eds., *Societies and Nature in the Sahel,* SEI Global Environment & Development Series (London: Routledge, 1997), 1–10. In particular, Raynaut calls the combined drought and subsequent Sahelian famine the quintessence of a major environmental disaster.

9. Diana K. Davis, *The Arid Lands* (Cambridge, Mass.: MIT Press, 2016), 160–63.

10. For more on the simplification of ecology through forestry, see James Scott, *Seeing like a State: How Certain Schemes to Improve the Human Condition Have Failed* (New Haven, Conn.: Yale University Press, 1998); and Donald Worster, *Nature's Economy: A History of Ecological Ideas* (Cambridge: Cambridge University Press, 1995).

11. F. Kenneth Hare, "Connections between Climate and Desertification," *Environmental Conservation* 4, no. 2 (1977): 81–90.

12. John A. Mabbutt, "The Impact of Desertification as Revealed by Mapping," *Environmental Conservation* 5, no. 1 (1978): 45–56.

13. David Hamilton Shinn and Joshua Eisenman, *China and Africa: A Century of Engagement* (Philadelphia: University of Pennsylvania Press, 2012), 52.

14. Shinn and Eisenman, 49.

15. The Hu Line is named for demographer Hu Huanyong. W. She, "Hu Huanyong: Father of China's Population Geography," *China Population Today* 15, no. 4 (1998): 20.

16. Wei Qi et al., "China's Different Spatial Patterns of Population Growth Based on the 'Hu Line,'" *Journal of Geographical Sciences* 26, no. 11 (2016): 1611–25.

17. A. K. Lobeck, *Geomorphology: An Introduction to the Study of Landscapes* (New York: McGraw-Hill, 1939), 381.

18. Armin K. Lobeck was a cartographer and graphic illustrator who integrated his global experience to account for the features of the landscape through geomorphic process. In the section "The Work of the Wind," he relates erosional features of wind to the transport of material as loess, confirming that the loess in China is two thousand feet deep in some areas. Lobeck, *Geomorphology,* 377–404.

19. David J. Mitchell and Michael A. Fullen, "Desertification and Reclamation in North-Central China," *Ambio* 23, no. 2 (1994): 131.

20. Ma Lifang et al., "Geology of China," in *Geological Atlas of China* (Beijing: Geological Publishing House, 2002), 9–12.

21. S. E. Smith, D. J. Read, and J. L. Harley, *Mycorrhizal Symbiosis* (Oxford: Elsevier Science, 1996).

22. Shixiong Cao, "Impact of China's Large-Scale Ecological Restoration Program on the Environment and Society in Arid and Semiarid Areas of China: Achievements, Problems, Synthesis, and Applications," *Critical Reviews in Environmental Science and Technology* 41, no. 4 (2011): 328.

23. See, for instance, Marijke Van Der Veen, "The Materiality of Plants: Plant–People Entanglements," *World Archaeology* 46, no. 5 (2014): 1–14; Jordan Goodman, Andrew Sherratt, and Paul E. Lovejoy, *Consuming Habits: Drugs in History and Anthropology* (Abingdon: Routledge, 2005); and Louis Lewin, *Phantastica: A Classic Survey on the Use and Abuse of Mind-Altering Plants* (Rochester, Vt.: Park Street Press, 1998).

24. Debra M. Shier and Ronald R. Swaisgood, "Fitness Costs of Neighborhood Disruption in Translocations of a Solitary Mammal," *Conservation Biology* 26, no. 1 (2012): 116–23. This study was the first empirical demonstration of the fitness consequences of disrupting social relationships among territorial neighbors.

25. Based on decades of failed survival attempts, the authors of a 2015 study claim that mitigation-driven translocations are key to the growing billion-dollar ecological consulting industry. Jennifer M. Germano et al., "Mitigation-Driven Translocations: Are We Moving Wildlife in the Right Direction?," *Frontiers in Ecology and the Environment* 13, no. 2 (2015): 100–105.

26. I am inspired by the work of Shirley Carol Strum as it relates to conservation, for her articulation of context over creature, preserving connective behavior over individuals. See Shirley C. Strum, *Almost Human: A Journey into the World of Baboons* (New York: Random House, 1987).

27. Assisted migration was first proposed with a deliberate emphasis on global warming by Robert L. Peters and Joan D. S. Darling, "The Greenhouse Effect and Nature Reserves," *BioScience* 35, no. 11 (1985): 707–17. Reference to assisted migration due to human-induced climate change is relatively new. See I. Aubin et al., "Why We Disagree about Assisted Migration: Ethical Implications of a Key Debate Regarding the Future of Canada's Forests," *Forestry Chronicle* 87, no. 6 (December 2011): 755–65.

28. John H. Pedlar et al., "Implementation of Assisted Migration in Canadian Forests," *Forestry Chronicle* 87, no. 6 (2011): 766–77.

29. The ectomycorrhizal *Pisolithus tinctorius* is typically inoculated in a variety of woody plants, while the genus *Glomus* is the most widely used endomycorrhizal fungi or vesicular-arbuscular mycorrhizal.

30. Lake States Forest Experiment Station, U.S. Forest Service, *Possibilities of Shelterbelt Planting in the Plains Region* (Washington, D.C.: Government Printing Office, 1935), 15.

31. A simple search for Chinese afforestation will yield coverage by popular publications rather than academic journals, including *Forbes, Wired,* and *National Geographic,* each using the sixty-six billion statistic without reference. This was a media claim by SFA. See, for instance, Alexandra E. Petri, "China's 'Great Green Wall' Fights Expanding Desert," *National Geographic,* April 21, 2007.

32. Nora Berrahmouni et al., "Building Africa's Great Green Wall: Restoring Degraded Drylands for Stronger and More Resilient Communities" (FAO, 2016), http://www.fao.org/3/a-i6476e.pdf.

8. THREE NORTHS SHELTER SYSTEM

1. Shixiong Cao et al., "Excessive Reliance on Afforestation in China's Arid and Semi-arid Regions: Lessons in Ecological Restoration," *Earth-Science Reviews* 104, no. 4 (2011): 240–45; *Bureau of Forestry Yearbook* (2006), cited in Shixiong Cao, Guosheng Wang, and Li Chen, "Assessing Effects of Afforestation Projects in China: Cao and Colleagues Reply," *Nature* 466, no. 7304 (2010): 315; J. J. Zhu et al., "Assessment of the World Largest Afforestation Program: Success, Failure, and Future Directions," *bioRxiv* preprint, February 3, 2017, http://dx.doi.org/10.1101/105619.

2. State Forestry Administration (SFA), *A Rising Green Wall: Construction of the Sanbei Shelter Forest System* (Beijing: Da Di Publishing, 1989), 16.

3. Zhu et al., "Assessment of the World Largest Afforestation Program."

4. According to a number of accounts and confirmed by the National Research and Development Center for Combating Desertification, Chinese Academy of Forestry. See, for instance, Xiaohui Yang, Longjun Ci, and Xinshi Zhang, "Dryland Characteristics and Its Optimized Eco-productive Paradigms for Sustainable Development in China," *Natural Resources Forum* 32, no. 3 (2008): 215–27.

5. Zhu et al., "Assessment of the World Largest Afforestation Program."

6. Duan Feng Quan, the horse breeder quoted here, was relocated within the Kubuqi Desert. He is quoted in an interview translated by Ian Teh, a photographer who spent considerable time documenting the forced transition.

7. China National Committee for the Implementation of the UNCCD, *China Country Paper to Combat Desertification* (Beijing: China Forestry Publishing House, 1997).

8. For instance, the United Nations Environmental Program (UNEP) awarded the SFA with a medal in 1987 for "the construction of the Shelter-forests in the North." State Forestry Administration, *A Rising Green Wall*.

9. Chunyang Li, Jarkko Koskela, and Olavi Luukkanen, "Protective Forest Systems in China: Current Status, Problems and Perspectives," *Ambio* 28, no. 4 (1999): 341–45.

10. Xin Rong Li et al., "Long-Term Effects of Revegetation on Soil Water Content of Sand Dunes in Arid Region of Northern China," *Journal of Arid Environments* 57, no. 1 (2004): 1–16.

11. The entire procedure is detailed in David J. Mitchell and Michael A. Fullen, "Desertification and Reclamation in North-Central China," *Ambio* 23, no. 2 (1994): 131–35.

12. Calculations include pedagogical description, accretion rates, and organic content. Further details are in Michael A. Fullen and John A. Catt, *Soil Management: Problems and Solutions* (London: Hodder Education, 2004), 40–45.

13. Conversation with representatives from Shapotou Desert Experimental Research Station, Northwest Institute of Eco-environment and Resources, Chinese Academy of Sciences, Lanzhou, China, November 2018.

14. State Forestry Administration, *A Rising Green Wall*, 29.

15. "The increased water shortage due to climate change could achieve 160 to 5090 million m^3 in some areas of China." Ying Aiwen, "Impact of Global Climate Change on China's Water Resources," *Environmental Monitoring and Assessment* 61, no. 1 (2000): 187.

16. This is a case study that is well translated and documents risk in the first phase, 1992–98. See, for instance, Sun Danfeng, Richard Dawson, and Li Baoguo, "Agricultural Causes of Desertification Risk in Minqin, China," *Journal of Environmental Management* 79, no. 4 (2006): 348–56.

17. Dennis Normile, "Getting at the Roots of Killer Dust Storms," *Science* 317, no. 5836 (2007): 316.

18. Sun, Dawson, and Li, "Agricultural Causes of Desertification Risk."

19. There are numerous reports that outline the particulars of each stage with the same vocabulary; the following one is specific to Shapotou: Ping Lü, Zhibao Dong, and Xiaoming Ma, "Aeolian Sand Transport above Three Desert Surfaces in Northern China with Different Characteristics (Shifting Sand, Straw Checkerboard, and Gravel): Field Observations," *Environmental Earth Sciences* 75, no. 7 (2016): 1–9.

20. Zhu et al., "Assessment of the World Largest Afforestation Program," 20.

21. Lü, Dong, and Ma, "Aeolian Sand Transport," 2.

22. Jian-jun Qu et al., "Study on Comprehensive Sand-Protecting Efficiency of Semi-buried Checkerboard Sand- Barriers," *Journal of Desert Research* 25, no. 3 (2005): 329–35.

23. Mitchell and Fullen, "Desertification and Reclamation in North-Central China."

24. Rudolf Geiger, *The Climate Near the Ground* (Cambridge, Mass.: Harvard University Press, 1957).

25. Li, Koskela, and Luukkanen, "Protective Forest Systems in China," 341.

26. Zhu et al., "Assessment of the World Largest Afforestation Program," 2.

27. In academic translations, the region is variously referred to as Horqin or Korqin, the same region.

28. According to a number of relevant reviews or books on desert dust outbreaks, dust particle size, and composition, of significance here are Joanna E. Bullard et al., "High-Latitude Dust in the Earth System," *Reviews of Geophysics* 54, no. 2 (2016): 447–85; Wang Xunming et al., "The Significance of Gobi Desert Surfaces for Dust Emissions in China; an Experimental Study," *Environmental Earth Sciences* 64, no. 4 (2011): 1039–50.

29. Qian Ma, "Appraisal of Tree Planting Options to Control Desertification: Experiences from the Three-North Shelterbelt Programme," *International Forestry Review* 6, nos. 3–4 (2004): 328.

30. FAO, "Tackling Desertification in the Korqin Sandy Lands through Integrated Afforestation," 2002, http://www.fao.org/docrep/006/AD115E/AD115E00.HTM.

31. FAO, "Tackling Desertification in Korqin."

32. MDGs stress measurable targets and clear deadlines for improving the lives of the world's poorest people. To meet these goals and eradicate poverty, leaders of 189 countries signed the historic millennium declaration at the United Nations Millennium Summit in 2000. "Video Message by Foreign Minister Wang Yi at the Launch Ceremony of the Report on China's Implementation of the Millennium Development Goals," *Beijing Review,* no. 34 (2015): I0007–8.

33. Shixiong Cao, "Why Large-Scale Afforestation Efforts in China Have Failed to Solve the Desertification Problem," *Environmental Science and Technology* 42, no. 6 (2008): 1826–31.

34. Cao, 1827.

35. Cao, Wang, and Chen, "Assessing Effects of Afforestation Projects in China."

36. See Thomas E. Barchyn et al., "Fundamental Mismatches between Measurements and Models in Aeolian Sediment Transport Prediction: The Role of Small-Scale Variability," *Aeolian Research* 15 (2014): 245–51.

37. Ma, "Appraisal of Tree Planting Options," 327–34.

38. Productivity increased particularly during the Han period, from 206 B.C. to A.D. 220. John L. Brooke, *Climate Change and the Course of Global History: A Rough Journey* (Cambridge: Cambridge University Press, 2014), 317–28.

39. Brooke, 319.

40. For an excellent overview of the rise of domesticated crops, see Daniel Zohary and Maria Hopf, *Domestication of Plants in the Old World: The Origin and Spread of Cultivated Plants in West Asia, Europe, and the Nile Valley* (Oxford: Oxford University Press, 2000).

41. There are numerous scholars who describe the simplification of plant resources. I rely on Alfred W. Crosby, *Ecological Imperialism: The Biological Expansion of Europe, 900–1900,* Studies in Environment and History (Cambridge: Cambridge University Press, 1986).

42. Isis Almeida and Megan Durisin, "American Wheat Great Again as Exports to Overtake Russia's," *Bloomberg,* May 25, 2017, https://www.bloomberg.com.

43. See, for instance, Jeremy Hegle, "Fighting Food Deserts in Kansas," Federal Reserve Bank of Kansas City, December 15, 2017, https://www.kansascityfed.org.

9. Species *Populus*

1. Scott Pauley, "Forest-Tree Genetics Research: *Populus* L.," *Economic Botany* 3, no. 3 (1949): 299–330.

2. A full list of *Populus* spp. (more than thirty-five records) is available from the USDA website, accessed October 1, 2021, https://plants.usda.gov/home.

3. Charles Darwin, *The Foundations of the Origin of Species: Two Essays Written in 1842 and 1844 by Charles Darwin,* ed. Francis Darwin (Cambridge: Cambridge University Press, 1909), 44; Lynn Margulis and Dorion Sagan, *What Is Life?* (New York: Simon & Schuster, 1995), 233.

4. Richard Karban uses the term *decisions.* See, for instance, Richard Karban, *Plant Sensing and Communication* (Chicago: University of Chicago Press, 2016), 78.

5. For a general discussion of vegetative reproduction, see Taylor A. Steeves, *Essentials of Developmental Plant Anatomy* (Oxford: Oxford University Press, 2016), 32–34.

6. Nicholas John Deacon et al., "Genetic, Morphological, and Spectral Characterization of Relictual Niobrara River Hybrid Aspens (Populus × Smithii)," *American Journal of Botany* 104, no. 12 (2017): 1878–90.

7. See Robert Ainsworth and Alexander Jamieson, *Latin Dictionary; Morell's Abridgment* (London: Moon, Boys & Graves, 1828), defined in two entries: Populus: People (5); Poplar tree (14).

8. For an excellent paleobotanical review, see Edward Wilber Berry, "Notes on the History of the Willows and Poplars," *Plant World* 20, no. 1 (1917): 16–28.

9. For an inspiring account of the plant as an individual, see Francis Hallé, *In Praise of Plants* (Portland, Ore: Timber Press, 2002), 113–21.

10. An excellent summary of sprouting is provided by Peter Del Tredici, "Sprouting in Temperate Trees: A Morphological and Ecological Review," *Botanical Review* 67, no. 2 (2001): 121–40.

11. Karban, *Plant Sensing and Communication,* 80.

12. Steeves, *Essentials of Developmental Plant Anatomy,* 126.

13. Michael J. Hutchings and Dushyantha K. Wijesinghe, "Performance of a Clonal Species in Patchy Environments: Effects of Environmental Context on Yield at Local and Whole-Plant Scales," *Evolutionary Ecology* 22, no. 3 (2008): 314.

14. See, for instance, Elbert L. Little, *Sixty Trees from Foreign Lands,* Agriculture Handbook 212 (Washington, D.C.: Forest Service, U.S. Dept. of Agriculture, 1961), 26.

15. Hallé, *In Praise of Plants,* 113.

16. Pauley, "Forest-Tree Genetics Research," 301.

17. R. B. Hall et al., "Commercial-Scale Vegetative Propagation of Aspens," *USDA Forest Service General Technical Report NC—North Central Forest Experiment Station,* no. 140 (1990): 211–19.

18. For instance, shoots were recorded emerging up to fifty meters from the parent plant. T. A. Spies and B. V. Barnes, "Morphological Analysis of *Populus Alba, Populus Grandidentata* and Their Natural Hybrids in Southeastern Michigan," *Silvae Genetica* 30, nos. 2/3 (1981): 102–6.

19. Anthony Trewavas, *Plant Behaviour and Intelligence* (Oxford: Oxford University Press, 2014), chapter 25, "Intelligence and Consciousness." Trewavas maintains that a record of past environments is deposited and stored in protein phosphorylation states, an activity of cell proliferation.

20. "Poplar was chosen as the first tree DNA sequence decoded because of its relatively compact genetic complement, some 50 times smaller than the genome of pine, making the poplar an ideal model system for trees." Department of Energy, "Populus trichocarpa v3.0," Joint Genome Institute (JGI), September 14, 2006, https://phytozome.jgi.doe.gov/pz/portal.html#!info?alias=Org_Ptrichocarpa.

21. G. A. Tuskan et al., "The Genome of Black Cottonwood, *Populus trichocarpa* (Torr.&Gray)," *Science* 313, no. 5793 (September 15, 2006): 1596–1604.

22. Tuskan et al., "The Genome of Black Cottonwood," 1599.

23. Huaxin Zhang, *Poplar Breeding, Cultivation Technologies and Extension in China*, APAFRI Publication Series 4 (Kuala Lumpur, Malaysia: APAFRI, 1999) 56, vii.

24. Xiyang Zhao et al., "Comparative Analysis of Growth and Photosynthetic Characteristics of (*Populus Simonii* × *P. Nigra*) × (*P. Nigra* × *P. Simonii*) Hybrid Clones of Different Ploidides," *PLoS ONE* 10, no. 4 (2015): E0119259.

25. Zhang, *Poplar Breeding*, vii.

26. UNESCO, "Taklimakan Desert—Populus euphratica Forests," January 29, 2010, https://whc.unesco.org/en/tentativelists/5532/.

27. Petra Lang et al., "Growth and Water Use of *Populus euphratica* Trees and Stands with Different Water Supply along the Tarim River, NW China," *Forest Ecology and Management* 380 (2016): 139–48.

28. Qun-Jie Zhang and Li-Zhi Gao, "The Complete Chloroplast Genome Sequence of Desert Poplar (*Populus euphratica*)," *Mitochondrial DNA Part A* 27, no. 1 (2016): 721–23.

29. Cao is one of the most vocal critics of the project, and while he does not attend to the species itself, he claims "implementation of large-scale afforestation has ignored the differences in topography, climate and hydrology, all of which affect tree survival." Shixiong Cao, "Why Large-Scale Afforestation Efforts in China Have Failed to Solve the Desertification Problem," *Environmental Science and Technology* 42, no. 6 (2008): 1826.

30. Cao, 1828.

31. Frederic E. Clements and John E. Weaver, *Plant Ecology* (New York: McGraw-Hill, 1938), 149.

32. Trewavas, *Plant Behaviour and Intelligence*, 273.

33. Ecology framed though competition begins with succession. See, for instance, early publications on plant succession, such as Frederic E. Clements, *Plant Succession: An Analysis of the Development of Vegetation* (Washington, D.C.: Carnegie Institution of Washington, 1916).

34. Clements and Weaver, *Plant Ecology*, 148.

35. H. Jochen Schenk, "Root Competition: Beyond Resource Depletion," *Journal of Ecology* 94, no. 4 (2006): 732.

36. Karban, *Plant Sensing and Communication*, 75.

37. Karban, 80.

38. J. Philip Grime, *Plant Strategies, Vegetation Processes, and Ecosystem Properties*, 2nd ed. (Chichester, West Sussex: Wiley, 2001).

39. A. J. Trewavas, "Profile of Anthony Trewavas," *Molecular Plant* 8, no. 3 (2015): 350.

40. Daniel Simberloff and Marcel Rejmánek, *Encyclopedia of Biological Invasions* (Berkeley: University of California Press, 2011), 158–59.

41. Michael J. Hutchings and Dushyantha K. Wijesinghe, "Performance of a Clonal Species in Patchy Environments: Effects of Environmental Context on Yield at Local and Whole-Plant Scales," *Evolutionary Ecology* 22 (2008): 315.

Index

Page numbers in italics indicate photographs or other illustrations.

ROSETTA S. ELKIN is associate professor and academic director of landscape architecture at Pratt Institute, principal of Practice Landscape, and research associate at the Arnold Arboretum of Harvard University. She is author of *Tiny Taxonomy: Individual Plants in Landscape Architecture.*